ASCE Manuals and Reports on Engineering Practice No. 46

Pipeline Route Selection for Rural and Cross-Country Pipelines

Prepared by the
Task Committee to rewrite Manual of Practice No. 46
Committee on Pipeline Installation and Location
Pipeline Division
American Society of Civil Engineers

Task Committee Members
Nicholas B. Day, Chairman
James A. Clark
Kenneth W. Kienow
Ralph C. Hughes
Peter K. Willerup

Published by

1801 Alexander Bell Drive
Reston, Virginia 20191-4400

Abstract: *Pipeline Route Selection for Rural and Cross-Country Pipelines* (ASCE Manuals and Reports on Engineering Practice No. 46) is a revision of the original manual No. 46, *Report on Pipeline Location,* published in 1965. Since that time, many high technology items have been developed to help the Routing Engineer, the Project Manager, and other project team members to be more productive in their jobs. The Task Committee on Pipeline Route Selection made an exhaustive review of information available on procedures that have been developed regarding the use of these new technologies. In addition to the discussion of high technology developments, this version of the manual places much more emphasis on the environmental, regulatory, and political issues related to pipeline route selection. Although safety and economic issues are addressed throughout the volume, a brief chapter on each topic explores some important considerations. Finally, a glossary of terms has been added since many terms may be unfamiliar or known under other names. This revision is aimed at all engineers involved in pipeline route selection. For some, it may be their first exposure to pipeline route selection considerations; for others, it may be a checklist of steps to go through to ensure a successful process.

Library of Congress Cataloging-in-Publication Data

Pipeline route selection for rural and cross-country pipelines / prepared by the task committee to rewrite manual of practice no. 46, Committee on Pipeline Installation and Location, Pipeline Division, American Society of Civil Engineers.
    p. cm.—(ASCE manuals and reports on engineering practice ; no. 46)
"Task committee members: Nicholas B. Day ... [et al.]."
Updates portions of: Report on pipeline location. 1965.
Includes bibliographical references and index.
ISBN 0-7844-0345-7
    1. Pipelines—Location—United States. 2. Network analysis (Planning) 3. Pipelines—Environmental aspects—United States. I. Day, Nicholas B. II. American Society of Civil Engineers. Committee on Pipeline Installation and Location. III. Report on pipeline location. IV. Series.
TA660.P55P58   1998
665.5'44—dc21   98-26052
                                                                                                         CIP

    The material presented in this publication has been prepared in accordance with generally recognized engineering principles and practices, and is for general information only. This information should not be used without first securing competent advice with respect to its suitability for any general or specific application.
    The contents of this publication are not intended to be and should not be construed to be a standard of the American Society of Civil Engineers (ASCE) and are not intended for use as a reference in purchase of specifications, contracts, regulations, statutes, or any other legal document.
    No reference made in this publication to any specific method, product, process, or service constitutes or implies an endorsement, recommendation, or warranty thereof by ASCE.
    ASCE makes no representation or warranty of any kind, whether express or implied, concerning the accuracy, completeness, suitability, or utility of any information, apparatus, product, or process discussed in this publication, and assumes no liability therefore.
    Anyone utilizing this information assumes all liability arising from such use, including but not limited to infringement of any patent or patents.

Photocopies: Authorization to photocopy material for internal or personal use under circumstances not falling within the fair use provisions of the Copyright Act is granted by ASCE to libraries and other users registered with the Copyright Clearance Center (CCC) Transactional Reporting Service, provided that the base fee of $8.00 per chapter plus $.50 per page is paid directly to CCC, 222 Rosewood Drive, Danvers, MA 01923. The identification for ASCE Books is 0-7844-0345-7/98/$8.00 + $.50 per page. Requests for special permission or bulk copying should be addressed to Permissions & Copyright Department, ASCE.

Copyright © 1998 by the American Society of Civil Engineers.
All Rights Reserved.
Library of Congress Catalog Card No: 98-26052
ISBN 0-7844-0345-7
Manufactured in the United States of America

MANUALS AND REPORTS
ON ENGINEERING PRACTICE

(As developed by the ASCE Technical Procedures Committee, July 1930, and revised March 1935, February 1962, and April 1982)

A manual or report in this series consists of an orderly presentation of facts on a particular subject, supplemented by an analysis of limitations and applications of these facts. It contains information useful to the average engineer in his everyday work, rather than the findings that may be useful only occasionally or rarely. It is not in any sense a "standard," however; nor is it so elementary or so conclusive as to provide a "rule of thumb" for non-engineers.

Furthermore, material in this series, in distinction from a paper (which expressed only one person's observations or opinions), is the work of a committee or group selected to assemble and express informaton on a specific topic. As often as practicable the committee is under the direction of one or more of the Technical Divisions and Councils, and the product evolved has been subjected to review by the Executive Committee of the Division or Council. As a step in the process of this review, proposed manuscripts are often brought before the members of the Technical Divisions and Councils for comment, which may serve as the basis for improvement. When published, each work shows the names of the committees by which it was compiled and indicates clearly the several processes through which it has passed in review, in order that its merit may be definitely understood.

In February 1962 (and revised in April 1982) the Board of Direction voted to establish:

> A series entitled "Manuals and Reports on Engineering Practice," to include the Manuals published and authorized to date, future Manuals of Professional Practice, and Reports on Engineering Practice. All such Manual or Report material of the Society would have been refereed in a manner approved by the Board Committee on Publications and would be bound, with applicable discussion, in books similar to past Manuals. Numbering would be consecutive and would be a continuation of present Manual numbers. In some cases of reports of joint committees, bypassing of Journal publications may be authorized.

MANUALS AND REPORTS OF ENGINEERING PRACTICE

No.	Title	No.	Title
13	Filtering Materials for Sewage Treatment Plants	69	Sulfide in Wastewater Collection and Treatment Systems
14	Accommodation of Utility Plant Within the Rights-of-Way of Urban Streets and Highways	70	Evapotranspiration and Irrigation Water Requirements
34	Definitions of Surveying and Associated Terms	71	Agricultural Salinity Assessment and Management
35	A List of Translations of Foreign Literature on Hydraulics	72	Design of Steel Transmission Pole Structures
37	Design and Construction of Sanitary and Storm Sewers	73	Quality in the Constructed Project: A Guide for Owners, Designers, and Constructors
40	Ground Water Management	74	Guidelines for Electrical Transmission Line Structural Loading
41	Plastic Design in Steel: A Guide and Commentary		
45	Consulting Engineering: A Guide for the Engagement of Engineering Services	75	Right-of-Way Surveying
		76	Design of Municipal Wastewater Treatment Plants
46	Pipeline Route Selection for Rural and Cross-Country Pipelines	77	Design and Construction of Urban Stormwater Management Systems
47	Selected Abstracts on Structural Applications of Plastics	78	Structural Fire Protection
		79	Steel Penstocks
49	Urban Planning Guide	80	Ship Channel Design
50	Planning and Design Guidelines for Small Craft Harbors	81	Guidelines for Cloud Seeding to Augment Precipitation
51	Survey of Current Structural Research	82	Odor Control in Wastewater Treatment Plants
52	Guide for the Design of Steel Transmission Towers	83	Environmental Site Investigation
		84	Mechanical Connections in Wood Structures
53	Criteria for Maintenance of Multilane Highways	85	Quality of Ground Water
54	Sedimentation Engineering	86	Operation and Maintenance of Ground Water Facilities
55	Guide to Employment Conditions for Civil Engineers	87	Urban Runoff Quality Manual
57	Management, Operation and Maintenance of Irrigation and Drainage Systems	88	Management of Water Treatment Plant Residuals
		89	Pipeline Crossings
59	Computer Pricing Practices	90	Guide to Structural Optimization
60	Gravity Sanitary Sewer Design and Construction	91	Design of Guyed Electrical Transmission Structures
62	Existing Sewer Evaluation and Rehabilitation	92	Manhole Inspection and Rehabilitation
63	Structural Plastics Design Manual	93	Crane Safety on Construction Sites
64	Manual on Engineering Surveying	94	Inland Navigation: Locks, Dams, and Channels
65	Construction Cost Control		
66	Structural Plastics Selection Manual	95	Urban Subsurface Drainage
67	Wind Tunnel Model Studies of Buildings and Structures		
68	Aeration: A Wastewater Treatment Process		

CONTENTS

HISTORICAL SUMMARY ix

FOREWORD xi

ACKNOWLEDGMENTS xvi

1 INTRODUCTION 1
    Responsibilities of the Routing Engineer 5
    Desirable Qualities and Knowledge of the Routing Engineer 5
    Planning Criteria and Tools 6
    Maps and Aerial Photographs 7
    Land Ownership Maps and Title Searches 7

2 ROUTE SELECTION 9
    Project Team 9
    Prerequisites to Pipeline Selection 9
        Products 9
        Terminal Points 10
        Delivery Points 10
        Pipe Size and Pipe Material 10
    Engineering Considerations 10
        Right-of-Way/Work Strip Width 10
        Access Road Rights-of-Way 11
        Public Roads 11
        Crossings—Water, Highway, and Railroad 11
        Trenchless Construction Techniques 15
        Land Uses 16
        Expanding Service 23
        Terrain 23
        Soil Conditions 25
        Unstable and Hazardous Locations 25
        Length 26

Above-Ground Pipelines..26
Other Considerations..26
Summary...27
Beneficial Routing Conditions....................................27
Routing Conditions To Be Avoided...............................27
Corridor Selection..28
Route Selection..31
Alignment Selection..35
Site Considerations..35
Conclusion..38

3 SAFETY..39
Introduction...39
Public Safety...39
Lifeline Safety..40
Mitigating Pipeline Incidents..40
Mitigating Pipeline Sabotage..42
Mitigating Incidents to Other Substructures (During Pipeline Installation)...43
Subsurface Utility Engineering..................................43
Existing Utility Records..43
Utilities' Information Quality Levels..............................45
Summary..46

4 REGULATORY AND POLITICAL ISSUES..............................47
Introduction...47
Agencies..49
Federal Agencies...49
State Agencies...49
Local Agencies...49
Local Political..50
National or Regional Special Interest Groups.......................50
Religious or Tribal Interest Groups...............................50
Local Special Interest Groups...................................50
The Public Involvement Program (PIP)................................50
Format of Meetings...53
Public Notification...53
Holding Meetings..54
Communications...55

5 ENVIRONMENTAL CONSIDERATIONS..............................57
Introduction...57
Purpose and Need..58
Scoping and Project-Related Studies.................................58
Environmental Compliance or Considerations.........................62
Corridor Studies...63
Routing Alternatives..64
Environmental Consequences.......................................64
Earth Resources..64

Biological Resources...64
Land Use Impacts...65
Socioeconomic Effects..65
Visual Impacts..65
Physical Impacts...65
Preferred Route Selection 65

6 ACQUISITION OF LAND RIGHTS....................................67
Securing Land Rights..67
Use of Franchises...69
Property Owner Relations..70

7 CONSTRUCTION, MAINTENANCE, AND OPERATION 73
Typical Construction Activities...73
  Mobilization...73
  Clearing of Work Strip..73
  Trenching..77
  Spoil Disposal 77
  Installing the Pipe...77
  Backfilling..79
  Strength Testing 79
  Cleanup and Work Strip Restoration 79
Typical Maintenance and Operation Activities...........................79

8 ECONOMICS 81
Pipeline Economic Considerations 81
Primary Economic Factors..82
  Estimated Cost of Building 82
  Estimated Annual Operating Expenses 82
  Forecasted Escalation Rate 82
  Assumed Value of Money...83
Secondary Economic Factors 83
  Economic Weather..83
  System Owner's Perception 83
Summary..83

GLOSSARY 85

BIBLIOGRAPHY AND REFERENCES 91

INDEX 93

HISTORICAL SUMMARY

The original task committee on Pipeline Location was formed in 1958 for the purpose of developing literature on surveying and other field engineering matters as they apply specifically to pipeline location. Nine reports appearing in the ASCE *Pipeline Division Journal* between February 1961 and January 1964 formed the substance of Manual 46, *Report on Pipeline Location*. Publication of the report was authorized by the Board of Direction on October 19, 1965.

In 1990, the Pipeline Installation and Location Committee formed several task committees to update Manual of Practice 46, and decided to divide the original manual into several different areas. This manual update covers route selection for pipelines that can be classified as rural or cross-country. Other task committees cover pipeline route selection in urban and other constricted areas, design installation, testing, and acceptance of pipelines.

The title for Manual of Practice 46 has also been changed from *Report on Pipeline Location* to *Pipeline Route Selection for Rural and Cross-Country Pipelines*. As a matter, perhaps of semantics, many people in the industry identify pipeline location with the finding of existing underground lines, and we wished to draw a clear distinction.

Nicholas B. Day was chairman of the Committee for this update, with the following additional members: James A. Clark, the late Ralph C. Hughes, Peter K. Willerup, and Kenneth W. Kienow. Pertinent information from the original Manual of Practice 46 has been updated and is included in the rewritten manual along with additional new information related to pipeline route selection.

FOREWORD

The history of pipelining is the story of a boom industry similar to the boom in the oil industry. The first major long cross-country pipelines were built by the oil refining companies to transport crude oil from the oil-producing areas of Oklahoma, Kansas, and western Texas to the areas of Kansas City, St. Louis, and Chicago. Although there had been short lines constructed in Pennsylvania and in the producing and refining areas of the Midwest and Southwest, it was the development of the pressure, electric, seamless, and continuous weld methods of pipeline construction in the late 1920s and 1930s that began the boom in cross-country transmission line construction.

In the late 1930s and 1940s, the natural gas transmission industry developed as producers in the gas fields of western Kansas, western Texas, New Mexico, eastern Texas, and the Gulf Coast area, together with bankers, promoters, and utilities in the metropolitan areas of the North Central and Northeastern states, raced to transport gas from wells to homes and industries. In the late 1940s and 1950s, thousands of miles of large-diameter natural gas transmission pipeline were constructed from these producing areas to loop the lines into the North Central and Eastern states and to build and extend new lines into population and industrial centers of the Pacific Northwest, New England, the California Coast, and Florida. Beginning in the 1930s, growing continuously through the 1940s and 1950s, and accelerating in the 1960s, the refined products' pipeline industry has grown along with the crude and natural gas transmission industry.

The economics of transporting by pipeline heralded a continuing rapid growth for the pipeline industry, not only for the oil industry, but also for the transportation of ever-increasing quantities of water and commodities that can be carried in fluid form.

The work of civil engineers and surveyors has been a major factor in the rapid growth of the pipeline industry. Considering the fact that there are hundreds of thousands of miles of pipelines in the ground, it is somewhat incongruous that there are very few textbooks and manuals on pipeline engineering and surveying, and that most of these publications have been prepared by pipeline companies. Except for a few articles appearing in technical journals, there still has been no authoritative book or manual since the appearance of the first edition of Manual 46 over 30 years ago.

Methods of performing preliminary centerline surveys for pipelines have changed drastically since the 1930s, whereas those for as-built surveys have changed little. Many of the first cross-country oil pipelines as well as some of the early natural gas and product pipelines were surveyed with compasses and line-of-sight methods of aligning laths and stakes. Costly meanderings across the terrain were of little concern to the early pipeliners, because the competitive economics of pipeline transportation were too advantageous.

Few pipelines were referenced to geodetic or State Plane Coordinate control systems, azimuth traverse control was inadequate, and aerial photographs were not available for projection of direct distance alignments between construction hazards. Land values were of minor consideration, rights-of-way being purchased for 25 cents per rod. Few titles were searched and plotted, and political boundaries, property lines, and ties were established infrequently. In those circumstances, the civil engineer had limited capacity or opportunity to design or locate routes that were best suited for construction, maintenance, and operation of pipelines.

However, as pipelines became more numerous they extended into areas of higher land values and diverse topography. Rights-of-way costs, and consequent needs for accurate and complete land surveys and title searches for property boundaries, became important; also, more attention had to be paid to safety, liability, and engineering problems.

The pipeline engineering field continues to expand. Natural gas use, especially, is on the rise as a source of inexpensive, abundant, and most important, clean energy. Deregulation has heated up competition, with new companies, or joint ventures between existing companies, forming and getting into the market at a fast pace. More and more engineers are being drawn into the industry, and professional requirements are high.

Today, the Routing Engineer (see the Glossary of Terms for this definition and for similarly used terms), the Project Manager, and other pipeline project team members, have at their disposal many high technology items to help in doing their jobs more productively. LandSat (NASA) or SPOT (French) imagery—high altitude color satellite photographs—color IR (infrared), color, and black and white photographs are available for preliminary corridor and route planning and for identification of land uses. Global Positioning Systems (GPS) can be used to determine accurately the location of the pipeline anywhere on earth within a few minutes. Aerial photogrammetry can provide accurate and detailed plan mapping of a centerline swath at almost any scale and width, together with a vertical profile. Digital orthophotography allows presentation of an accurate scaleable photomap that can act as a backdrop for detailed information such as property lines and other utilities. Together with field data gleaned from total stations (electronic theodolites/transits) and data collectors, information can be downloaded directly into a computer-aided drafting (CAD) system for interactive design of the pipeline alignment right on a workstation screen monitor. Geographic Information Systems (GIS) interfaced with CAD are also available for storing, manipulating, and querying all the information about a pipeline. This includes route constraints, land uses, pipeline as-built details, maintenance and operating information, environmental and property ownership details, gas customers and revenues—all tied together in a relational database. Pen-based laptop computers are the most recent tool to be used effectively in gathering, changing, or adding measurements and attributes right on site, the pen being used directly on the screen drawing. These technologies are available now, have been used successfully by several pipeline companies for recent projects, and should all be considered part of a company's arsenal when planning, constructing, and operating a new pipeline.

The Task Committee on Pipeline Route Selection made an exhaustive review of information available on procedures that have been developed and are in use by engineers throughout the pipeline industry. The members of the Committee represent the natural gas, crude oil, and products' pipeline industry, and work in all parts of the United States. In addition, they represent the surveying, construction engineering, engineering administration, and consulting engineering aspects of the industry.

The text of the original manual was devoted primarily to suggested procedures related to cross-country pipelines: reconnaissance, location, construction, and as-built records—examined with due attention to right-of-way requirements throughout. Special conditions for major water or river crossings, marsh and offshore locations, building sites, and microwave communications were treated, and several useful appendix items were also included in the original Manual 46.

A Bibliography and Reference section remains, detailing references used in this manual, and those that may prove useful for expanded knowledge. A Glossary of Terms has been added for this rewrite. Many terms may be unfamiliar, or known under other names, depending on the background of the reader, and this glossary should ensure that the text is fully understood.

The format of the 1965 manual has also been changed somewhat by eliminating the detailed descriptions of surveying procedures in the reconnaissance, preconstruction, and as-built survey phases. They are considered only incidental to the route selection process, and are dealt with here in as much detail as is felt necessary to support the process. The reader should refer to the numerous books on engineering surveys, including ASCE Manuals 34 and 64, for additional information, and ASCE Manual 75, *Right-Of-Way Surveying*.

In addition, this manual update places much more emphasis on the environmental, regulatory, and political issues related to pipeline route selection. Two new chapters—Regulatory and Political Issues, and Environmental Considerations—have been added to cover these crucial areas. Ignoring them, or simply paying them lip service, especially when trying to route a new pipeline in a highly regulated or environmentally concerned state will jeopardize a project's schedule and add considerably to overall costs. All states are subject to the same federal laws, regulations, and acts affecting the pipeline industry, many of which have been enacted since the original Manual of Practice was published in 1965, but states and counties vary widely in their own laws, ordinances, and regulations. Some states have their own public utilities commissions (PUCs) that regulate *intrastate* pipelines, whereas other states may have no PUC, but still retain some sort of review process. Major *interstate* transmission pipelines are regulated by the Federal Energy Regulatory Commission (FERC) and the Department of Transportation (DOT).

Included within the chapter on Regulatory and Political Issues is a section on public involvement. This, again, is a vitally important process that must be started early on in a project. In the public's mind, we are no longer opening up new frontiers with new pipelines. Much of the time they want to preserve the status quo, and are suspicious of the need for new projects. Experience or long memories of past projects, the information age, and the activity of special interest groups have made the general public very knowledgeable about many aspects of pipeline projects. They want to know what's going on, why, where, when, and how. They want good communication. They want to be asked their opinions and be treated honestly and fairly, without patronization. Too much information is too readily available to too many people today for pipeline companies to be anything but completely up front with project details. Project proponents must win public support to be successful. To have a win–win situation they must develop a credible strategy and be prepared for a good amount of give and take, or compromise. No matter how competent the routing engineer, design engineer, or construction engineer feel, and how important or costly the materials and construction might be, they are all irrelevant if the public and/or regulators do not accept the project.

Finally, two brief chapters have been included on safety and economics. Although these subjects pop up throughout the manual as they relate to route selection, the text in these chapters covers some important considerations. References for more in-depth reading are also mentioned.

As to restricting content, the Committee felt this manual should only cover procedures, methodologies, and issues commonly found in the route selection of proposed pipelines in the continental United States. Even though there are many references to oil, petroleum products, and natural gas pipelines, the underlying principles of route selection are valid for all types of large diameter, high-pressure, cross-country pipelines such as water, wastewater, and slurry products. Also, many of the steps, caveats, and ideas described in this manual can be applied or modified to special conditions encountered in the location of offshore, Arctic, or tropical pipelines.

This rewritten manual is directed to all engineers involved in pipeline route selection. For some, it may be their first exposure to pipeline route selection considerations; for others it may be a checklist of steps to go through to ensure a successful process. Or, it may be a reminder that, for some pipelines, there are many more issues and problems than those of the specific pipeline on which the routing engineer is currently working.

As a final note, and in consideration of ASCE's encouragement for all engineers and surveyors to adopt Systeme Internationale units, all measurements throughout this manual are metric. However, realizing that many engineers and surveyors are still working in the Imperial system, we have shown equivalents in parentheses. Exact conversions have not always been attempted as distances mentioned in the manual are often approximate guidelines. Soft and hard conversions are used as appropriate.

ACKNOWLEDGMENTS

The committee responsible for rewriting Manual of Practice No. 46 would like to thank the following people for reviewing the manual drafts. Their insightful comments, experience in pipelining, and additions to the text have helped us produce what we hope will be useful industry guidelines. They represent both the private and public sectors, from the West Coast to the East Coast, and overseas.

James H. Anspach, Allan T. Bartz, Bruce A. Bennett, Richard N. Davee, Karen J. Dawson, Brian J. Doeing, Richard W. Gailing, Andrew Jenkins, Loren C. Kolditz, William R. Ledbetter,Jr., A.C.J. McLaughlin, Walter R. McLean, R.C. Prevost, William F. Quinn, Ted Shettler, Malcolm Stephens, John J. Struzziery, and Dan Williams.

Chapter 1

INTRODUCTION

Organizing a manual such as this into chapters that follow specific steps in a logical order is difficult. Although the process flows from the initial planning phase through to pipeline construction, it has become so highly integrated that a logical schematic diagram is somewhat unrealistic. However, Figure 1-1 shows part of a typical project management schedule outlining some of the basic tasks.

The routing engineer, surveyor, environmental planner, design engineer, right-of-way (ROW) agent, construction engineer, and so on, cannot work independently in a vacuum at successive stages of the project to complete their particular assignments. Figure 1-2 shows a typical company organization chart that may be put in place for a pipeline project. Figure 1-3 depicts, in more detail, the hub and spoke approach, or perhaps more accurately, the spider's web concept of a totally integrated team approach where virtually all players interact with each other. There is much exchange of information in what is an iterative process. The chapters are set out in as logical an order as possible, but should not be considered reflective of the actual flow of events. Certain chapters overlap in their details and descriptions so that each is self-sufficient to a certain extent and shows how each item fits into the big picture; several important concepts and issues are repeated. If there is one main objective or concept that the reader should come away with after reading this manual, it is an appreciation for the "big picture." Of necessity, the reader will have to refer back and forth in the manual for issues and concerns that may occur throughout the project.

The words routing engineer, corridor, route, and alignment, are found throughout the manual. Although these terms are noted in the Glossary of Terms at the back, they form such an integral part of the routing process that we first define them here:

- Routing Engineer: the person who has the pivotal position in any project involving the routing of new pipelines.

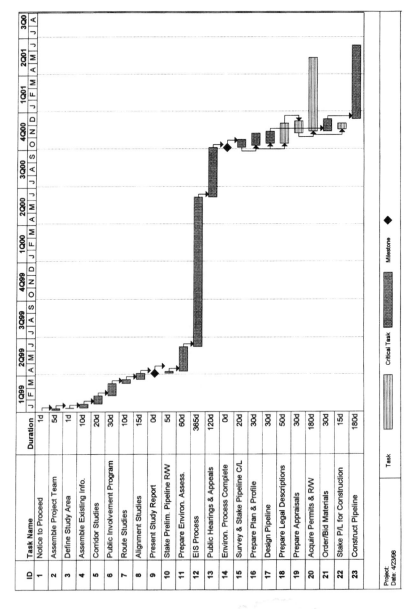

FIGURE 1-1. *Sample Pipeline Project Management Schedule.*

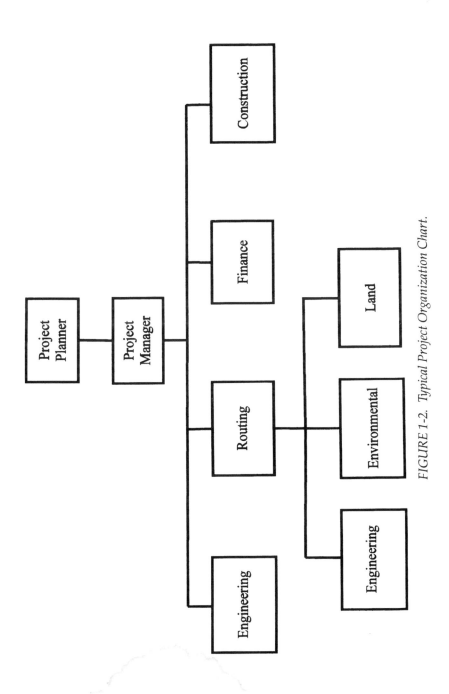

FIGURE 1-2. *Typical Project Organization Chart.*

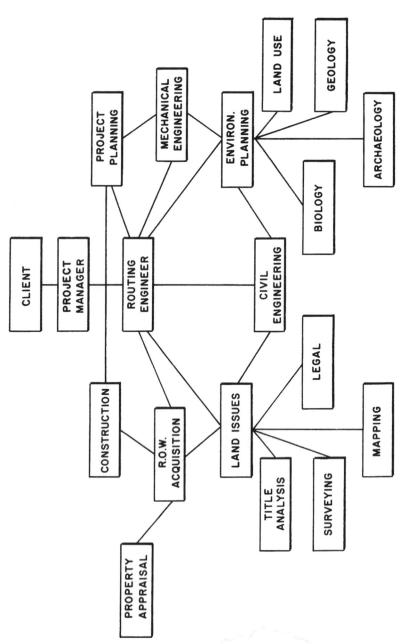

FIGURE 1-3. *Project Team Relationship.*

- Corridor: a strip of land of variable width, often one to three kilometers ($1/2$ to 2 miles) wide, that accommodates, or could accommodate, one or more utility facilities such as a pipeline.
- Route: a strip of land of varying width, typically 300 to 400 meters (1000 feet to $1/4$ mile) wide, within a corridor, and within which a right-of-way for a pipeline could be located.
- Alignment: the actual location of the pipeline; the centerline of the proposed facility as established by a survey.
- Right-of-Way: the actual width of land, usually purchased as an easement rather than in fee, required to safely maintain and operate the pipeline and protect it from future development. An additional width is often acquired temporarily for a construction working strip.

RESPONSIBILITIES OF THE ROUTING ENGINEER

The routing engineer is the person chosen to have the final decision on the routing of a new pipeline alignment. After considerable input from other members of the project team, the routing engineer will make a recommendation to the project proponent on the preferred and alternative alignments before they are presented to the regulators, various agencies, and public. As discussed later, there are many different aspects to consider and balance before making the final decision.

The routing engineer must be given the authority and flexibility by the project manager to make the compromises necessary to reach that final decision. The reason for this sole responsibility is the necessity of having one person objectively weighing the various special interests of the project proponents (i.e., the project planner, project manager, customer, etc.), the environmental constraints, the public's interests, and the political aspects.

The routing engineer is also the person who usually defends the location in public meetings, public hearings, in front of granting authorities, and in court if required. With an intimate knowledge of most issues, and having spent considerable time on the ground, the routing engineer will exact the most credibility with outside entities. The routing engineer will not normally be expected to defend the need for the project or the proposed costs and rate requirements.

DESIRABLE QUALITIES AND KNOWLEDGE OF THE ROUTING ENGINEER

Routing a pipeline should be considered as much an art as a science. There is no specific or standard way to route a pipeline. All projects are different and present their own problems and solutions, but all involve the coordination of large amounts of data from many sources and numerous contributors.

Here are some ideal requirements.

- Sufficient knowledge to understand the engineering planning that leads up to the proposed project, and ability to interact with the planners in suggesting possible alternatives to the proposed project.
- Working knowledge of pipeline design and construction engineering. Experience shows that if the routing engineer is not a design or construction engineer, those engineers will be the closest advisors.
- An awareness of what is needed to perform maintenance on the pipeline for the duration of its useful life.
- Intimate understanding and compassion for the environment and its experts. Consider their advice and findings, which may often result in a longer route and/or more costly project. No one expertise has priority!
- Political awareness and knowledge of the laws and regulations that govern the study area, and ability to apply these constraints during the routing process. The routing engineer must know who the political decision makers are—those politicians, powerful landowners, and special interest groups—who may support or oppose the project.
- A demeanor and presence that allows him or her to explain the proposed route to individuals and public groups, in informal or official surroundings, and to accept objections and criticism without losing composure.
- Open-minded and ready to accept input from diverse sources. This will allow a balanced objective decision in routing the proposed pipeline.
- Meticulous preparation that will make him or her feel comfortable in defending that location as the preferred route in the judicial courts, if necessary.

The expert routing engineer acquires this skill by experience and education, and retains this expertise by an ongoing awareness of all the changes in engineering, environmental sciences, economics, and public politics.

PLANNING CRITERIA AND TOOLS

At one time, the main criterion for line location was engineering design; but real estate costs, environmental impacts, and construction efficiency now significantly affect the route selection process. Therefore, several routes should be selected tentatively, with costs and design developed for the various alternatives. Not only is this usually required now by most pipeline companies' management, but is mandatory for compliance with certain states' environmental review processes [e.g., the California Environmental Quality Act (CEQA) requires equal evaluation of at least two alternative routes], together with an in-depth analysis of the alternatives. For pipelines falling under the jurisdiction of the Federal Government a review is required under the National Environmental Policy Act (NEPA). The areas of real estate, environment, and construction are covered in greater detail in the following chapters.

MAPS AND AERIAL PHOTOGRAPHS

To facilitate planning and make the best informed decision, the routing engineer must get the most up-to-date information available about the study area. Good corridor planning requires detailed maps and photographs; a vast amount of these already exist within the United States. The US Geological Survey (USGS) publishes topographic maps at scales of 1:24,000, 1:62,500, 1:250,000, and 1:500,000, and the Bureau of Land Management (BLM) 1:100,000 planning maps. The Defense Mapping Agency (DMA), US Army Corps of Engineers (USCOE), National Ocean Survey (NOS), and Tennessee Valley Authority (TVA) can also provide useful maps. Other useful information can be obtained from various federal, state, and local agencies.

Many areas are covered by existing aerial photography at various scales. Federal and state agencies (notably the US Department of Agriculture) as well as many photogrammetric mapping companies, have extensive film libraries that often cover remote areas. A number of private companies have a regular flight coverage program that depends on the extent of new or anticipated development within their territory. Other sources of reconnaissance photography that can be ordered in black and white, regular color, or infrared color (IR) are NASA, SPOT Image (French; available at distributors in the USA), and Russian satellites; however, the scale of some of the imagery may not be adequate for many projects. If no adequate and recent photography exists, new flights of the study area or selected corridors should be considered since it will be useful throughout the project. Because one wide corridor may contain several possible routes, the type of photography required will depend on whether several corridors are to be covered or several routes in one corridor.

LAND OWNERSHIP MAPS AND TITLE SEARCHES

The importance of land ownership or assessors' maps is often underestimated when doing route studies. These should be acquired early in the process. The number of owners and type of ownership along an alignment can be critical factors in the final selection.

The preliminary route alignments should be plotted on the ownership maps as soon as they are known. This initiates the search for property owner names, which will be needed for notification of public meetings. Also, an early search of legal land documents can turn up easements that may preclude location of a pipeline, and which may not be physically observable on the ground. This could include land owned by railroads or irrigation districts where no tracks or canals currently exist, but are held for future use.

Naturally, some of the first items to check when selecting a route are the existence of road and utility rights-of-way and other easements, which may be exclusive. This may often be noted on ownership or assessors' maps. However, usually a check must be made with the various responsible entities, such as public works' departments, city and county planning departments, and private and public utility companies.

Chapter 2
ROUTE SELECTION

PROJECT TEAM

The routing engineer is, as described in Chapter 1, the person responsible for the final location of a proposed pipeline. The routing engineer should consult with many other experts and form a team at the project's outset to narrow proposed alternatives into a final pipeline alignment. The composition of the team depends on the anticipated obstacles and concerns that may be found in the area of the proposed project.

The team should include engineers with expertise in the fields of underground transmission pipelines, their design, and the mechanical and construction methods. It should also include specialists in land acquisition and environmental planning. Again, depending on the area or terrain traversed by the proposed transmission line, the team may expand to include governmental affairs representatives and attorneys. If the environmental concerns are anticipated to be high, the environmental planner may seek the advice of specialists such as biologists (both terrestrial and marine), geotechnical engineers or geologists, and cultural resource and agricultural experts. Please refer again to Figure 1-3 which shows a typical team relationship for a major pipeline project.

PREREQUISITES TO PIPELINE SELECTION

Before the route selection process can begin, there are many facts that need to be known. These are often determined by transmission planners, engineers, market forces, customers, and company standards.

Products

First, what type of product(s) has to be transported? This often sets the parameters for engineering constraints, and routing of the proposed pipeline.

Terminal Points

These must be known, but if not absolute, at least the approximate area should be known. For example, the proposed pipeline will tap into an existing line, but the transmission planners will allow this point anywhere along a certain segment.

Delivery Points

Controlling points along the proposed pipeline need to be known. There may be tap or delivery points along the route that must be reached, or reasons other than just pipeline location for placing valve, meter, or compressor stations.

Pipe Size and Pipe Material

These should be known as they normally set many design and construction requirements. They are described in detail in Chapter 7. Pipe size and material may constrain horizontal and vertical curves, or the type of construction equipment that will be required, together with access and width of the construction impact zone. It should also be known if, and how, tunneling or horizontal boring are contemplated. This will be important in locating water, highway, or railroad crossings.

ENGINEERING CONSIDERATIONS

The routing engineer must use the expertise of many others in the areas of pipeline design and planning, construction, operation, and maintenance to assist in selecting the most appropriate pipeline route for the given set of circumstances. The intent here is to point out some of the engineering issues that should be considered when selecting a pipeline route.

Right-of-Way/Work Strip Width

The width needed for a pipeline right-of-way, or work strip, is partly dependent on the size and type of pipeline to be constructed (see Figure 2-1 and Table 2-1). The right-of-way should be wide enough to allow for normal operation and maintenance (including replacement) of the pipeline considering the potential for development outside and adjacent to the right-of-way. The width of the work strip (temporary land rights used during construction) should be such that the most economical construction methods can be used during pipeline installation.

For more discussion on types of easements and right-of-way considerations, please refer to ASCE's Manual of Practice 75, *Right-of-Way Surveying*.

Access Road Rights-of-Way

Opportunities to parallel public or private roads are another routing advantage. A route requiring minimal new road construction for access reduces costs and environmental impacts (see Figure 2-2). For further information on the types of rights to be acquired, again refer to Chapter 6 and ASCE Manual 75 on *Right-of-Way Surveying*.

Public Roads

Public roads are considered advantageous for transporting construction workers and materials. However, construction within the city, county, or state highway system may be less desirable than constructing on private rights-of-way for the following reasons.

- High installation costs may occur due to more difficult construction, spoil removal, and repaving (subject, of course, to terrain).
- Relocation at the pipeline owner's expense may be required for future road improvements where the installed pipeline interferes with road reconstruction. There may also be franchise or permit conditions, in addition to throughput fees. Relocation, of course, may depend on the specifics of a franchise agreement and/or who has senior rights.
- Construction in roadway franchise can be less desirable since lane closures and traffic management are required. The potential hazard created by lane closures during construction increases risks to the motoring public and workers. Limitations on the amount of lane closure allowed during nonworking periods requires barricading, signing, and suitable steel plate covering of the open trench.
- Disturbance to the road shoulder and pavement by pipeline construction carries the potential for future claims by the city, county, and/or state for any subsequent damage to the roadway. Damages are generally paid by the pipeline owner.
- Roadway or roadside excavation requires the removal and end haul of substantial volumes of material taken from the pipeline trench. Road bed grade material must be imported and compacted to the road authority's density standard. Haulaway, of course, might be necessary in open country as well.

Some of these same disadvantages can apply to railroads and canals, or other linear facilities.

Crossings—Water, Highway, and Railroad

When looking at a potential pipeline route, water bodies, highways, and railroads can be viewed as control points. In many cases, suitable crossing locations of these features are selected and the rest of the pipeline fit

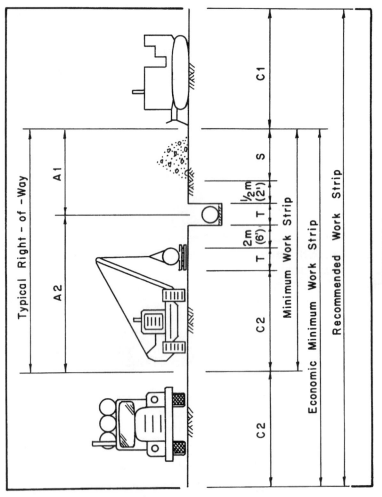

FIGURE 2-1.

TABLE 2-1. Guidelines for Construction Work Strip Widths/Right-of-Way for Various Size Pipe on Level Terrain.

Pipe Size	Pipe 2(T)	Trench Setback	Spoil Pile S	Dozer Length C1	Sideboom Width C2	2C2	Work Strip			Right-of-Way				Construction Equipment
							Minimum 2(T)+2.5 +S+C2	Economic Min. 2(T)+2.5+ S+2C2	Recommended 2(T)+2.5+S +C1+2C2	Typical Permanent Right-of-Way	Typical Pipe Offset A1=S+0.5 +0.5(T)	Typical Pipe Offset A2=C2+2 +1.5(T)		
150mm (6")	900mm (3')	2.5m (8')	1.5m (5')	4.9m (16')	4.3m (14')	8.5m (28')	9.1m (30')	13.4m (44')	18m (60')	9-11m (30-36')	2.4m (8')	6.7m (22')	D-4 & Whirley
200-300mm (8"-12")	1.2m (4')	2.5m (8')	1.8m (6')	5.5m (18')	5.5m (18')	11m (36')	11m (36')	16.5m (54')	22m (72')	11-14m (36-45')	2.7m (9')	8.2m (27')	D-6 & 561 Sideboom
400-600mm (16"-24")	1.8m (6')	2.5m (8')	3.6m (12')	6.4m (21')	5.8m (19')	11.6m (38')	13.7m (45')	19.5m (64')	26m (85')	14-17m (45-55')	4.6m (15')	9.1m (30')	D-7 & 572 Sideboom
660-900mm (26"-36")	3m (10')	2.5m (8')	5.2m (17')	6.7m (22')	6.1m (20')	12.2m (40')	16.8m (55')	22.9m (75')	30m (97')	17-30m (55-100')	6.4m (21')	10.4m (34')	D-8 & 583 Sideboom

Notes:
(1) These figures are for use with pipe sizes from 150mm (6") to 900mm (36") and level construction areas where soil and site conditions, local regulations, permits, weather, etc. allow standard trenching and pipeline installation methods. Other conditions must be evaluated on a case-by-case basis.
(2) *Recommended* is the desired work strip width for all locations where space is available.
(3) *Economic Minimum* is the minimum work strip in which a pipeline could be economically installed under normal conditions.
(4) *Minimum* is the minimum work strip required to install the pipeline. Installation within this work width will substantially decrease productivity and increase costs (for less than *Minimum*, see note (5)).
(5) Where practical, the *Right-of-Way Width* should be equivalent to the *Minimum* work strip width. *Right-of-Way* widths can be reduced for some conditions (e.g., the ROW is adjacent to existing easements, along a roadway, etc.). A reduced ROW width must allow for normal operation and maintenance (including replacement) of the pipeline considering the potential for development on one or both sides of the ROW.

FIGURE 2-2. Existing Farm Road Used for Access Route C/L Flagged to Avoid Well.

between them. Crossings deserve special consideration, especially those involving water bodies.

Pipeline routes may cross rivers, streams, and washes or run parallel to a waterway. Typically, pipes are buried in the stream bed or elevated above on bridge structures. In either case, there is a risk that the pipeline will be washed out or exposed during major floods. Hazards to the pipe include: vertical scour, bank erosion, channel meandering, and bridge scour. If these areas cannot be avoided, a hydrologic and erosion analysis should be conducted to determine the design discharge, recommended burial depth, and lateral limits of scour. As an alternative to deep burial, grade control or bank stabilization techniques may be evaluated. In addition, refinements can be made to the pipeline alignment to minimize scour hazards.

High river banks, which require extensive cut and spoil removal, should be avoided. Crossing on a curve is also to be avoided, where possible, as it contributes to the scour and erosion mentioned above. Environmental concerns can include disturbance of endangered species habitat and the potential for geyser effects. Levees, banks, and approaches may have to be replaced, contributing to the need for greater working space.

Depending on the approaches, terrain, and available working space, crossings should be as close to perpendicular as possible. These are best from both a permit grantor and grantee standpoint. Minimum length of exposure is best and eases identification of the pipeline location using markers. Most companies limit the crossing angle to an absolute maximum of 45°. No side bends (PIs) should be permitted within 50' to 100' of the crossing limits to allow uniform straight-run pipe installation, whether cased or uncased. For railroads, there must be sufficient room to bore at a depth that will not undermine the tracks.

Construction experts and environmental permitting personnel should be involved in any major crossing selection. When crossing wetlands areas, the project team must weigh the benefits and costs of conventional trenching versus a horizontal directional bore (see below). However, a comparison can only be made for limited lengths.

Trenchless Construction Techniques

At this point, we should mention the phenomenal growth in recent years in horizontal directional drilling, trenchless technology, and pipe bursting and their potential impact on route selection in certain areas. Eyed with suspicion in the early days by permitting agencies for crossing rivers and avoiding disturbance to endangered species and levees, horizontal directional drilling has gained considerable respectability. In fact, it appears to be their preferred choice.

Many areas, once cost-prohibitive to cross with a pipeline, are now more feasible through the use of trenchless techniques such as horizontal directional drilling (HDD) (see Figure 2-3); slick boring, bore and jack (see Figures 2-4 and 2-5); microtunneling, aerial (bridging) (see Figure 2-6); and a number of pipeline rehabilitation methods. The big selling point of these trenchless techniques is minimal surface disturbance. This translates to (1) no broken pavement or traffic disruptions; (2) no waterway disturbance to interfere with fisheries, water quality, or shipping; and (3) no disturbance to endangered or historical structures, plants, or wildlife. The cost for this type of pipeline construction continues to decrease as the technology is used more frequently and is more eagerly accepted by permitting agencies.

In selecting a route with the intention of using HDD, the routing engineer must ensure that there is sufficient room on each side of a crossing for pipe stringing and laydown area, entry and exit points, and bore pits.

The routing engineer needs to be aware of the various trenchless techniques available for pipeline installations, and the advantages of each, so that sensitive area crossings may be presented to the design engineer as viable, alternative pipeline routes.

There are many publications that go into great detail on the various trenchless pipeline construction techniques. A good reference is the ASCE Pipeline Division, Manual Of Practice (MOP) 89, *Pipeline Crossings*, 1996. Additional reference material regarding this subject can be found in ASCE Pipeline Division Conference Proceedings, *Pipeline Crossings*, 1996 and *Trenchless Pipeline Projects—Practical Applications*, 1997.

Land Uses

Grazing/Rural Undeveloped. Pipelines are highly compatible with most extensive land uses. Disturbances to these lands during construction are temporary. After construction, normal land use can resume with few restrictions. Obvious exceptions are where there are rare and endangered species, or cultural resources. It should also be noted that there are certain soil conditions and geology where the construction scar may remain for many years.

Agricultural—Dry/Non-Irrigated. Pipelines are compatible with dry land farming. Where possible, construction should be scheduled after crops are harvested. It is common practice to compensate growers for crop losses.

Agricultural—Intensive/Irrigated/Orchard/Vineyard. Pipelines are compatible with most intensive agricultural uses, with certain caveats. Where possible, alignments are chosen to follow existing farm roads, irrigation routes, or crop rows. Deeper placement of the pipeline (1.5 meters or more (60 inches)) is necessary where deep ripping is a normal farming practice. Pipelines in orchards or vineyards should attempt to parallel existing rows or irrigation systems where possible. Pipelines located in orchards containing deep rooted trees are generally not compatible.

FIGURE 2-3. Existing Pipeline in Hard Shoulder and Under Slough (Butte County, CA).

FIGURE 2-4. Bore Pit for Road Crossing.

ROUTE SELECTION

FIGURE 2-5. Bore and Jack Equipment on Sled.

FIGURE 2-6. Colorado River at Topock, CA/AZ Border; Foreground: Old Route 66 Bridge Now Used to Carry Pipelines; Background: Suspended Pipeline from Towers.

ROUTE SELECTION

Note: There will be occasions when a company wants to parallel one of their existing pipelines, perhaps using the same right-of-way. However, when the original pipeline was constructed it may have been across virgin land, but where row crops, orchards, or housing have now grown up. Crossing at the head of a land-leveled field, growing tomatoes or lettuces, for example, and with an intricate system of irrigation pipes, can be a costly and disruptive process. Locating at the low point in the field, away from irrigation pipe systems and the existing pipeline, will win more advocates among the farming community. In this situation, if the right-of-way is wide enough, and has legal rights for multiple lines, exchanging those rights for a more suitable location within the same property may be considered (see Figures 2-7 and 2-8).

Residential. Pipelines are compatible with residential areas. However, a higher level of design criteria is usually needed to meet permitting and/or applicable code requirements. These are usually more restrictive in residential areas because the permitting agency is highly influenced by the property owners and can be very restrictive and untimely in their deliberations.

Future Residential. Pipelines can be compatible with future residential development areas if planned appropriately. The pipeline right-of-way may be used for green belts, open space, vehicle parking, roads, or recreation

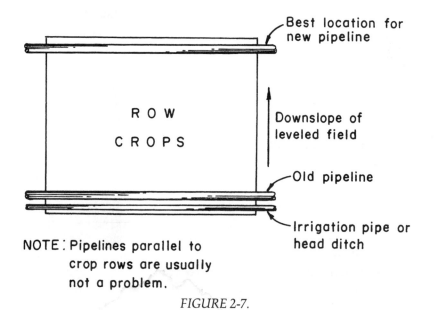

FIGURE 2-7.

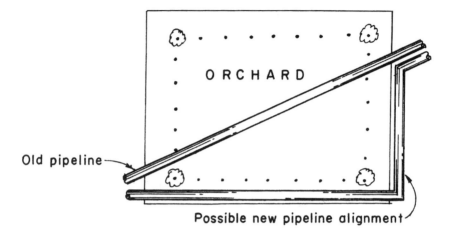

NOTE: Decision to avoid parallel of old pipeline will depend on economics and negotiations with property owner.

FIGURE 2-8.

paths as long as future use does not infringe upon the operation and safety requirements of the pipeline.

Industrial. Pipelines are compatible with industrial areas and may be located in street franchise or existing pipeline right-of-way. A caution for a possible issue within industrial areas is contaminated soil. If it is known before construction, the issue can be dealt with, either by removing the contaminated soil, or rerouting the pipeline. However, if the condition is found during construction there could be major delays in the schedule, with an ensuing cost increase.

Park/Open Space/Recreational. Again, pipelines are compatible with these types of land uses. However, the level of design criteria may be increased depending on the number of users, and as prescribed by applicable codes.

Reservoirs and Water Storage Areas. Pipelines are generally not compatible with reservoirs or lakes. Difficulties in pipeline maintenance, inspection, and repair are the major reasons for this incompatibility. Pipeline routes should avoid crossing reservoirs, unless the longer route causes excessive present and future economic impacts. Even if the pipeline is only paralleling a water storage area, rather than crossing it, there are certain considerations.

Most agencies responsible for the watershed area will not allow oil pipelines to be routed on the downhill side of the catchment area where a rupture could cause leakage into the water supply (see Figure 2-9).

Forests and Wooded Terrain. Pipelines can be compatible with forests or wooded terrain, but may require expensive removal of trees and their root systems for the pipeline alignment. Additional removal of vegetation may be required during the construction phase. Of course, the costs of circumventing a forest with longer pipe length must be weighed against those of the most direct route. Regulated forest areas, or commercial timberlands often have severe restrictions on where pipelines can be located. A cleared right-of-way through the forest serves as a fire break, and can act as a change in the environment for some wildlife. However, the corridor should not intersect a highway perpendicularly so as to be visible.

Mineral Deposits. Compatibility is dependent on type and location of the deposits. Areas where minerals are close to the surface, and are likely to be quarried or strip mined, should be avoided. It is unlikely that a pipeline company could ever obtain an easement here from the minerals rights owners. Oil, natural gas, water, or geothermal steam are a different matter. Obviously, as long as an existing wellhead is not within the right-of-way, these resources are too deep to interfere with a pipeline. Future exploration can usually be accommodated by slant drilling to the deposits from outside the right-of-way.

Expanding Service

Consideration should be given to selecting pipeline routes that can accommodate service to new customers not currently being served. However, speculative development plans that are not supported by approved plans should not be given serious consideration. The cost for this type of additional pipeline must be justified by revenues from these new customers.

Terrain

Pipeline routes must consider the operation of construction equipment and the handling of pipe and other pipeline components. Side-slope conditions should be avoided or minimized due to the requirement for excessive cut and fill during pipeline construction (see Figure 2-10). Where routing opportunities exist in similar terrain, a profile analysis should be performed to select the route with the least grade and elevation changes. In extremely rugged terrain, construction allowances must be made for

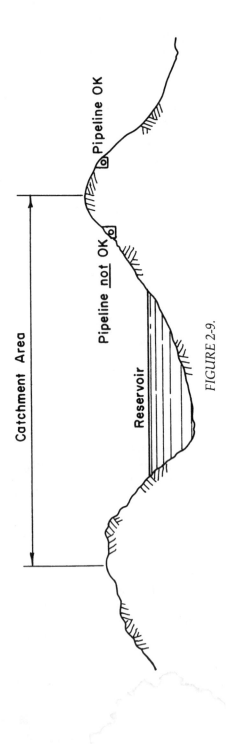

FIGURE 2-9.

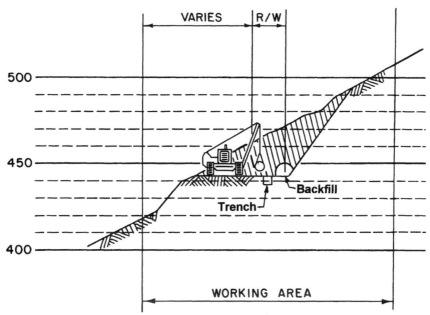

FIGURE 2-10. *Sideslope Conditions on the Existing Right-of-Way Require a Wider Construction Work Area to Accommodate Pipeline Equipment.*

steep pitches and changes of grade. Construction practices in steep terrain must take into consideration potential erosion and land stability problems that do not exist in level terrain.

Soil Conditions

Certain types of soils are compatible with pipeline construction methods on a seasonal basis only, and some soils produce "soil stress" on the pipeline and corrosion protection coating. Permafrost can require special construction and maintenance procedures; bedrock requires difficult construction techniques. A pipeline needs soil with good bearing capacity, not peats and certain muds that are subject to liquefaction.

Unstable and Hazardous Locations

Geological, seismic, and other unstable locations or hazards should be avoided for pipeline routes, if possible. When a long pipeline route is required, it is generally not possible to avoid crossing all the unstable locations that can be encountered. Although engineering design can accommodate crossings of potential hazards, site-specific, costly designs are often needed for each location. Routes that reduce the amount of exposure are preferable.

Length

Pipeline routes of greater length than other alternatives are subject to additional construction costs and time, increased exposure to dig-ins, increased operation and maintenance costs, and greater capital costs for pipe and right-of-way. Shorter routes are normally preferable from an environmental, engineering, and economic perspective. A "unit cost" per mile for material and construction is often a useful tool for evaluating the feasibility of increasing the length to avoid an obstacle.

Above-Ground Pipelines

There are a number of advantages and disadvantages for above-ground pipelines. Most companies have very strong views on this subject and strict policies, usually not to have any exposed pipe. Often, small diameter pipes, especially oil lines over oil companies' lands can be seen lying on the surface. All size lines, transporting a variety of products, can be seen strung beside or under bridges, across nonnavigable canals and aqueducts, suspended across canyons, or unsupported over streams and ditches. In some cases, there is no alternative to above-ground pipelines, and/or the cost of burying the line is exorbitant (e.g., the Alyeska Pipeline). The advantages can be cost savings in pipe length, design, and construction, time savings in installation, and easier access for maintenance. The disadvantages are the pipe's exposure to the elements, accidents, shooting sprees, and sabotage. In a number of cases, the new trenchless methods, directional drilling, and other technologies, combined with environmental issues, have made undergrounding more attractive. For a more in-depth treatise on this subject and crossings in general, please refer to ASCE's Manual of Practice 89, *Pipeline Crossings*, 1996.

Other Considerations

The route selected by the routing engineer should consider safety of the line (see Chapter 3, Safety), regulatory and political concerns (see Chapter 4, Regulatory and Political Issues), and environmental concerns (see Chapter 5, Environmental Considerations). To help route selection, the routing engineer should understand the construction, maintenance, and operation activities required for the proposed pipeline (see Chapter 7, Construction, Maintenance, and Operation). The route selected must have been compared to alternative routes using an economic analysis evaluation (see Chapter 8, Economics).

ROUTE SELECTION

Summary

Engineering considerations dictate that a particular condition or characteristic is either beneficial to a proposed pipeline route or should be avoided. Here is a checklist of typical conditions and characteristics that normally affect the selection of a pipeline route with "Beneficial" and "To Be Avoided" categories. These should help the routing engineer evaluate the alternatives.

Beneficial Routing Conditions

The routing engineer should look for pipeline route alternatives that favor these items.

- Existing roads paralleling the proposed route.
- Public roads available to transport workers, materials, and equipment to the work site.
- Proposed pipeline route is through or within:
  — grazing and rural undeveloped lands,
  — agricultural—intensive/irrigated/orchard/vineyard (in most cases; see also under "Routing Conditions To Be Avoided"),
  — agricultural—dry/nonirrigated lands,
  — a future residential area (only for a gas distribution system),
  — level to near-level terrain, or
  — an area of limited vegetation.
- Proposed pipeline route can accommodate expanded service to new customers (again only if a gas distribution system is anticipated).
- Pipeline can be installed in an environment of relatively constant and mild temperature. Of course, this is not always realistic, for example, Alaska or Saudi Arabia.
- Acceptable landfill and other required disposal sites near the proposed route.
- Acceptable pipe laydown and material and equipment storage sites along the proposed route.
- Proposed route can accommodate future pipeline maintenance and operation requirements.
- Proposed route is the shortest alternative, because construction and operational costs usually outweigh all other costs (subject, of course, to any mitigation costs, which can be substantial, and high-priced real estate).

Routing Conditions To Be Avoided

When an item(s) in this category cannot be avoided, the routing engineer should consult with the design and/or construction engineer to develop an appropriate solution, including cost and schedule impacts, to mitigate the negative effect of the particular condition or characteristic. This is done to facilitate a valid cost/benefit comparison with other alternatives.

- Public roads, highways, or freeways.
- Orchards containing deep-rooted species.
- Residential areas.
- Park, open space, and/or recreational lands.
- Unstable/hazardous locations:
  — earthquake fault,
  — flood and/or washout,
  — landslide,
  — pile driving,
  — sinkholes or karst topography,
  — soil liquefaction,
  — soil settlement, subsidence, or erosion, or
  — unusual overburden, wheel, and/or track loading.
- Rugged terrain and side-slopes.
- Habitat of protected rare and endangered or threatened species.
- Areas containing cultural resources.
- Other environmentally significant areas.
- Areas where general plans, policies, and/or ordinances are in conflict with pipeline construction, operation, and/or maintenance.
- Physical obstructions:
  — airport, airstrip, or helipad,
  — canal, river, creek, or stream,
  — hard rock requiring blasting,
  — high population density or building,
  — lake, pond, reservoir, or water storage,
  — levee,
  — major highway and railroad,
  — marsh, swamp, or wetland,
  — rock outcrop, or
  — substructure congested area.
- Mineral deposit areas subject to mining, particularly strip mining and quarries.
- Corrosive or contaminated soil.
- Proximity to high-voltage transmission lines and related facilities (AC induction problems, cathodic protection interference, etc.).

CORRIDOR SELECTION

When the controlling points have been selected, the routing engineer is faced with selecting the final pipeline alignment having the least impact on the area traversed, the least costs in construction and operation, and which meets the operational timeframe. These criteria alone may eliminate many alternative locations.

The first process is the selection of alternative corridors (see Figure 2-11) in which the proposed pipeline can be constructed. To select corridors, the routing engineer must know the major constraints within the study area (see Figure 2-12). These are engineering, environmental, and economic con-

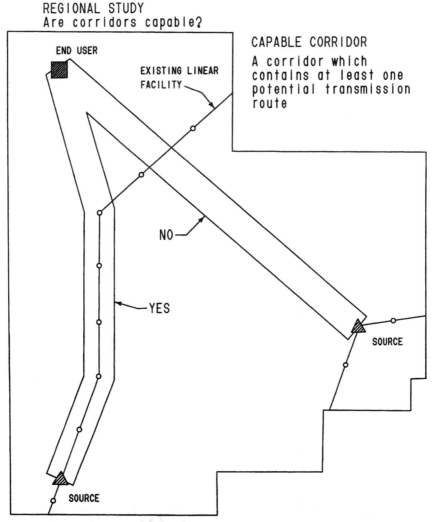

FIGURE 2-11. *Selection of Alternative Corridors.*

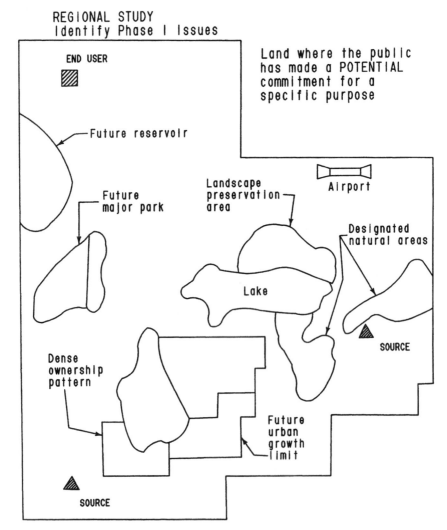

FIGURE 2-12. *Identification and Mapping of Constraint Areas (Phase I).*

straints, and areas of public concern and human development. They are described in detail in Chapter 5.

The routing engineer must also know the location of existing and proposed linear facilities with an eye to paralleling or jointly using the land rights and construction access. Linear facilities include other pipelines, electric transmission facilities, roads, railroads, and canals.

All major constraints and opportunities are best visualized when mapped on a topographical base map and used in conjunction with recent aerial photographs. A useful photographic scale for preliminary studies is 1:24,000, which matches that of $7^1/_2'$ USGS quad sheets, although 1:12,000 is better for picking out details in more congested areas. An increasing amount of information on constraints—endangered species, geology, cultural resources, and the like—is now available, often in national or local agency GIS databases. The routing team is then ready to select alternative corridors.

With modern technology, as much as 90% of all pipeline reconnaissance can be accomplished by map studies. In some cases, the selection of corridors can be done without a field inspection, but most often this is needed. It can be done from the air, either by fixed-wing airplane or helicopter, and in some cases by traveling on the ground. Aerial photographs will be of great assistance during this process.

Alternative corridors are linear strips of land of varying width depending on the length of the proposed project; for long lines they may range from 1 to 3 kilometers ($^1/_2$ to 2 miles) wide. The corridors must all be able to contain at least one acceptable route for the proposed pipeline, and are selected on the basis of best meeting the criteria of economics, engineering, environmental, and public concerns.

Once the alternative corridors have been selected, the major concerns and constraints should be described and/or mapped, in consultation with other team members, for each of the proposed corridors. These concerns and constraints may require an adjustment of the corridors (see Figure 2-13), but finally the routing engineer and project team will decide on the preferred corridor and selection of alternative corridors.

ROUTE SELECTION

The selection process now focuses on selecting alternative routes within the preferred corridor. A route is a strip of land of varying width, seldom wider than 400 meters ($^1/_4$ mile), that contains at least one acceptable alignment for the proposed pipeline. These routes will be identified and the team members will gather detailed information within the routes, evaluate constraints or opportunities (see Figure 2-14), adjust the alternative routes if needed, and select one preferred route and alternatives (see Figure 2-15).

The public puts a high value on these processes. In major projects, it is necessary to consult with public entities such as city, county, state, or federal planners to get their comments and input at this stage of the process. In many cases, informal meetings must be held where the public is invited to verify or add to the constraints identified by the project team. Official public hearings could also be required, which may result in an acceptance or denial of the proposed project and its routing. This subject is covered in greater detail in Chapters 4 and 5.

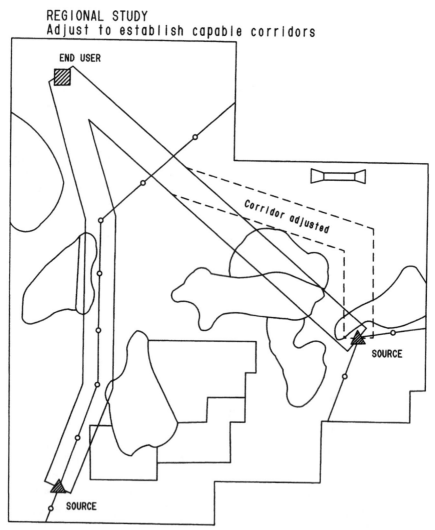

FIGURE 2-13. *Adjustment of Alternative Corridors.*

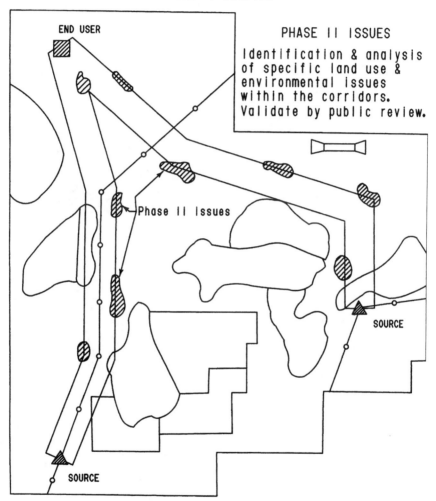

FIGURE 2-14. Corridor Analysis.

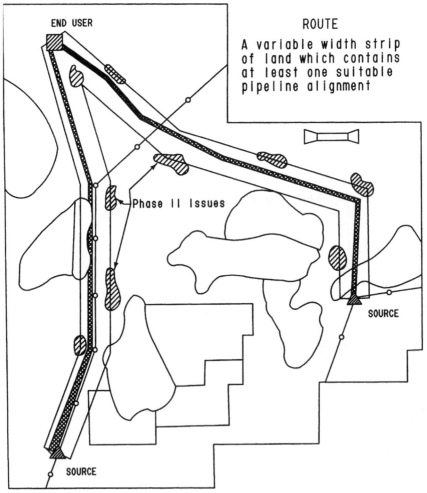

FIGURE 2-15. *Route Identification.*

Ranking of the routes and selection of the preferred route can be done using various techniques. One, using the matrix approach, which covers such items as length, right-of-way costs, environmental issues, economics, construction problems, and land uses, can be seen in Table 2-2. The routing engineer can make the decision independently, or the selection can be done by a team of experts who rank the proposed alternatives and come to a unified decision. In the case of major projects, subject to state or federal regulations, the lead agency will make the decision after input from the project proponent, the public, and the various agencies.

ALIGNMENT SELECTION

Within each of the proposed routes, or selected route, a final alignment will be chosen (see Figure 2-16). The alignment is the determination of the exact location of the proposed pipeline. This includes the selection of angle points, or curves, of the proposed pipeline alignment. At this point, most members of the team should have walked the entire route selected.

This sets the parameters for the preliminary centerline survey. From this, the final engineering design will be made, the material will be ordered, and land rights or permits will be mapped, described, and acquired.

SITE CONSIDERATIONS

The selection of the route and final alignment also includes the selection of locations for the ancillary facilities to the proposed pipeline. These facilities must be specified at the outset of the project by the engineering planner.

Compressor stations, pump stations, mainline valves, and "smart pig" receiving stations are normally facilities that are required for larger or longer pipelines. These sites are usually fenced-in areas, with equipment above ground, and will require land areas that are generally flat and easily accessible. After the general location and size have been specified by the engineering planner, the final selection of the site is made using criteria similar to the location of the pipeline itself.

The construction impact is of relatively short duration, but the station facilities will be in operation as long as the pipeline is in operation. The selection should therefore take into consideration access to the facility on a year-round basis as well as the environmental impact (noise, visual, etc.) that operation of the site will have on adjacent land use. Availability of electricity, water, and communication facilities are also factors in site selection.

Minor facilities, which may also be required for the operation of the pipeline, are regulator and dehydration stations, and crossovers.

TABLE 2-2. Route Alternatives—Sample Evaluation Criteria Matrix Analysis

Evaluation Criteria	Alternative Routes and Variations			
	Rte 1	Rte 1A	Rte 2	Rte 3
Engineering				
Route Length (km)	48.0	48.5	45.9	52.0
Parallel Positions (km)	9.0	9.4	5.1	16.2
Use of Franchise (km)	9.8	10.4	7.4	13.4
ROW Acquisition (km)	35.2	34.6	36.8	35.5
ROW Acquisition Costs ($000s)	922	930	895	763
Property Owners Crossed (No.)	51	41	39	47
Parcels Crossed (No.)	65	56	49	67
Major Water Crossings (No.)	5	5	4	3
Roads/Highways Crossed (No.)	3	3	4	4
Railroads Crossed (No.)	7	8	9	5
Faults Crossed (No.)	1	1	2	2
Major U/G Utilities Crossed (No.)	41	46	32	55
Pipe & Construction Costs ($000s)	15,000	15,200	14,300	16,300
Environmental				
Known/Proposed Development (km)	0.8	0.8	1.8	0.5
Future Urban Use (km)	1.1	1.1	0.8	0.8
Ag/Permanent Crops (km)	19.4	19.7	17	21.4
Ag/Annual Crops (km)	14.2	13.9	11.4	14.6
Using Existing Corridors (km)	16.5	15.8	11.7	17.9
Sensitive Archaeologic Areas (km)	1.9	1.9	1.9	1.4
Plant Habitat Areas (km)	5.3	5.3	5.3	6.2
Animal Habitat Areas (km)	7.7	7.7	8.2	6.2
Areas of Critical Concern (km)	0.3	0.3	0.5	0.2
Reclamation Districts (No.)	6	6	5	4
County & City Plan Areas (No.)	3	3	3	4
Forests, Parks (km)	1.8	2.2	3.8	0.3

Note: Depending on project, location, or situation, some criteria may be more critical than others, and so more weight should be applied. The engineering, environmental, or any other evaluation criterion, can be reduced or expanded as necessary.

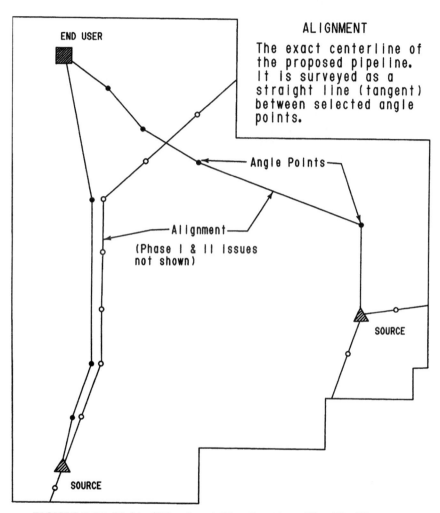

FIGURE 2-16. *Right-of-Way Acquisition: Location of Specific Alignment.*

All these facilities are different from the pipeline itself as they are above ground and require periodic access. They therefore create an impact on the land that is very different from the pipeline. Here, the routing engineer must make the same balanced decision as for the pipeline, narrowing down the location from larger land areas to the final site.

CONCLUSION

If all the steps described previously are followed, the routing engineer is assured that the proposed pipeline alignment is in a location where all factors have been considered, weighed, and understood, where the public has been part of the process, and where it has the least impact on the public and greatest value to the company. The process also ensures the minimum of objections to the project and potential delays during the design and construction phases.

Chapter 3
SAFETY

INTRODUCTION

Regardless of the type of commodity or product that is being transported by a cross-country pipeline, pipeline safety is basically divided into two areas of concern: public safety and lifeline safety.

Public Safety

This portion of pipeline safety is concerned with pipeline design, construction, operating and maintenance criteria, and procedures that have to do with resistance to damage that could directly result in injury, death, and/or significant property damage. The most critical element of pipeline public safety (life and property protection) is usually population density; that is, the greater the number of people located close to the pipeline, the greater the opportunity for third-party damage to occur.

Pipeline accident statistics have repeatedly shown that third-party encroachment is the most frequent cause of pipeline accidents, which subsequently result in injury, death, public or private property damage, and damage to the pipeline facility. There are numerous company, local, state, and federal damage prevention programs, such as one-call notification systems, intended to reduce the potential for third-party damage on operating pipelines. However, route selection for new pipelines and corridors offers an opportunity to take into consideration the potential for population density increases, thereby avoiding such future problem areas.

These considerations for the effect of population density, and the remedial routing of the pipeline, are a necessity regardless of the type of commodity or product being transported. The interruption of service due to third-party damage can be just as serious with any product contained in the pipeline.

In the case of natural gas and petroleum products, the minimum safety standards are mandated by federal and state regulatory agencies and are subject to audit and enforcement. Good references for this type of mandate are the US Department of Transportation's *Pipeline Safety Regulations—For Natural Gas and Hazardous Liquids Pipeline.*

Other publications that address life and property protection are industry standards and manufacturers' recommendations. An excellent overall reference is the Transportation Research Board (TRB) National Research Council's *Special Report 219—Pipelines and Public Safety,* published in 1988.

Lifeline Safety

This portion of pipeline safety is concerned with the critical systems of the overall pipeline delivery system and their effect on the system's ability to deliver. Pipeline lifeline safety looks at damage and failure by how it affects the system. Under normal conditions, this is mainly a pipeline owner's issue, such as loss of income, and usually takes a back seat to pipeline public safety. It is important to note that failure to deliver pipeline products critical to life (e.g., fuel, water) for extended periods in disaster situations (e.g., flood, fire, earthquake, hurricane) can become a public safety issue by indirectly resulting in injury, death, and/or property damage. The critical element of pipeline lifeline safety (serviceability protection) is closeness to the supply source.

MITIGATING PIPELINE INCIDENTS

The pipeline routing engineer needs to be aware that external forces, third party damage (dig-ins) and corrosion are the major causes of pipeline incidents in the United States. Future pipeline incidents can be mitigated to some extent through good pipeline routing practices. Here is a list of some routing considerations and mitigating practices, to avoid pipeline incidents, that should be included in any evaluation of route alternatives.

- Avoid paralleling high-voltage power lines with a metallic pipeline. If paralleling a high-voltage power line is needed as a pipeline route alternative, include a mitigation plan to eliminate AC current.
- Avoid locating a metallic pipeline close to DC power sources, such as light-rail transportation systems. If locating a pipeline near a DC power source is needed as a pipeline route alternative, include a corrosion mitigation plan that addresses large amounts of stray DC.
- Avoid farming areas where deep ripping is done. If this is not possible, ensure that adequate cover—at least 7 feet—is specified (see Figure 3-1).

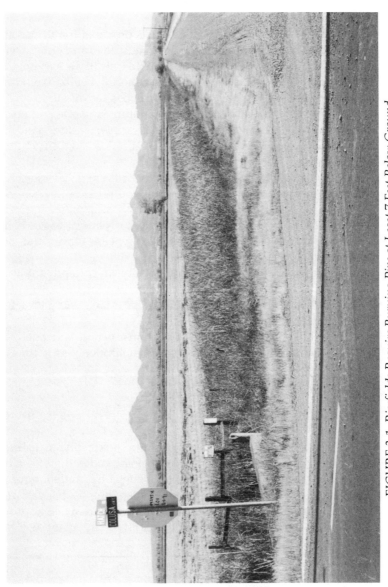

FIGURE 3-1. Ricefields Require Burying Pipe at Least 7 Feet Below Ground.

- Avoid corrosive soils. Obtain soil survey information when available and include a corrosion control plan in the route alternative evaluation process.
- Keep all above-ground natural gas and hazardous liquid piping away from overhead sources of ignition.
- Ensure that good engineering judgment is provided for all crossings (unstable soil, highways, waterways, railroads, environmentally sensitive areas, etc.). Design to last for the economic life of the system. Consider larger wall thickness, larger diameter, more cover than minimum requirements, and use of trenchless construction techniques.
- Obtain the most current information for future land use and ensure that the pipeline design accommodates known future land use changes. Accommodate any other known proposed substructures if accurate location information is available.
- Site stations with good drainage and access. Select pad elevations high enough to prevent flooding of the station. Sites should be remote enough to provide a barrier to vehicle traffic.
- Where possible, install the pipeline in an exclusive right-of-way. Land documents should be written to protect pipeline cover and limit encroachment.
- If the pipeline is designated as critical, specify cover depth and clearances from other substructures that exceed minimum requirements.
- Always try to locate the pipeline route where it is easy to patrol, both by land and air.
- Avoid side-hill construction. If side-hill construction is needed as a pipeline route alternative, include a slope stabilization plan in the evaluation.

MITIGATING PIPELINE SABOTAGE

Currently, pipeline sabotage is not a contributor to pipeline incidents in the United States; however, changing world politics and the effect of a "world economy" on the US workforce could change this situation for the worse. The pipeline routing engineer needs to keep the possibility of sabotage in mind and take mitigating measures along the route when appropriate. In addition to the preceding routing practices, the following will help mitigate pipeline sabotage.

- Limit any above-ground installation.
- Provide adequate property for stations to provide a good buffer distance between any above-ground facility and the security fence.
- Stations should be remote but within the view of property owners near the pipeline and the general public passing by.

MITIGATING INCIDENTS TO OTHER SUBSTRUCTURES (DURING PIPELINE INSTALLATION)

Subsurface Utility Engineering

There is an increasing impact on projects from existing underground utilities (see Figure 3-2). These existing utilities are becoming more prevalent as our population grows and our infrastructure ages and must be replaced. As time goes on, records or knowledge of them are becoming increasingly irretrievable or difficult to reference. Reliance solely upon known utility owner's records creates risks. These risks manifest themselves in utility relocation costs, extra work orders, change orders, construction and redesign delays, and potential damages to utilities with their resultant consequences of death, injury, property damage, and environmental releases. A pipeline routing engineer needs to have knowledge of subsurface utility engineering principles and practices. Subsurface utility engineering uses knowledge of surface geophysical methods, surveying, and engineering in order to classify the quality level (QL) of existing utility information on plans and construction documents. This allows one to identify and mitigate the risks for damage to a third party during the pipeline installation.

Existing Utility Records

The first place to start investigating for the locations of other substructures that are within a proposed pipeline route is existing records. In doing this it is very important to realize that reliance solely upon the existing records of utility owners can create problems.

- Records were not accurate in the first place—design drawings are often not "as-built," or installations were "field run," and no record was ever made of actual locations.
- On old industrial sites there have usually been several utility owners, architectural and/or engineering firms, and contractors installing facilities and burying objects in the area for decades. The records seldom get put in a single file and are often lost; there is almost never a composite.
- References are frequently lost; the records show something, say, 28 feet from a building that is no longer there, or from the edge of a two-lane road that is now four-lane or part of a parking lot.
- Lines, pipes, and tanks are abandoned, but do not get removed the drawings.
- Even so-called "as-builts" frequently lack the detail and veracity needed for design purposes in a utility-congested environment.

FIGURE 3-2. *Open Cut Pipeline Trench in Road Pavement Showing Exposed Subsurface Utilities and Shoring.*

Perhaps the largest problem encountered by both the design engineer and the constructor is that they do not know the quality of the utility information shown on plans. Information is typically always disclaimed by the engineer in a statement on the plans, such as:

> Utilities depicted on these plans are from utility owners' records. The actual locations of utilities may be different. Utilities may exist that are not shown on these plans. It is the responsibility of the contractor at time of construction to identify, verify, and safely expose the utilities on this project.

This statement exists whether utility information is known to be correct, or whether it is just information from a dubious verbal source.

Utilities' Information Quality Levels

Four distinct quality levels of utility information have been identified and endorsed by a variety of public and private agencies, such as the FHWA, NTSB, or USDOT. Utility quality levels are depicted on plans using any variety of techniques such as CADD levels, line codes, line weights, and symbols. They are defined as follows.

- *Quality Level "D"—Existing Records:* utilities are plotted from review of available records.
- *Quality Level "C"—Surface Visible Feature Survey:* QL "D" information from existing records is correlated with surveyed surface-visible features.
- *Quality Level "B"—Designating:* two-dimensional horizontal mapping. This information is obtained through the application and interpretation of surface geophysical methods. Utility indications are referenced to appropriate survey control.
- *Quality Level "A"—Locating:* three-dimensional mapping and other characterization data; this information is obtained through exposing utilities by test holes and measuring and recording (to appropriate survey control) utility and/or environment data, or recording these data during construction to recoverable survey control.

Quality level "D" is usually sufficient for the selection of route alternatives, but should be upgraded when preliminary and final design is under way.

Additional information on safety, subsurface utility engineering, and related subjects can be found in the Bibliography and References section.

SUMMARY

The routing engineer is very important in helping to avoid future pipeline incidents by providing the safest pipeline route possible. Both public safety and lifeline safety need to be addressed throughout the pipeline routing process. The major pipeline safety concerns that the routing engineer must accommodate are: external forces, third-party damage (dig-ins), corrosion, and sabotage.

The list of routing practices given in this chapter to assist with improving pipeline safety during the pipeline routing process is by no means complete. The routing engineer should use the list as a starting point and make a personal list of practices that is continually updated, expanded, and tailored to the type of conditions normally encountered.

To mitigate damage (dig-ins) to the structures of third parties during the installation of a proposed pipeline, the routing engineer needs to provide a good evaluation of other substructures that are within a proposed pipeline route. Also, information on the location and physical size of other identifiable substructures should be assigned a quality level of information accuracy. This will help the designer and constructor decide whether to use the existing information "as is," or take appropriate action to improve its accuracy.

Chapter 4

REGULATORY AND POLITICAL ISSUES

INTRODUCTION

For at least the past 10 to 15 years the regulatory aspect has controlled major projects, but in many cases, insufficient thought and consideration is given to both the regulatory and political processes when selecting pipeline routes. These are probably pre-eminent today in the success of a pipeline project. A seemingly insignificant issue can invalidate or delay an environmental impact statement (EIS) or report (EIR), or close down construction until resolved. Surveys, mapping, design, ordering materials, and construction are fairly straightforward activities where most procedures and variables are predictable. Nothing is predictable when dealing with regulatory agencies and the public.

Regulatory and political considerations, or issues, are considered together here in the same chapter because, so often, they are closely interwoven in the overall acceptance and construction of a pipeline. However, in a pure sense, the two differ in that anything regulatory must conform to certain written laws, regulations, ordinances, and policies. These could be federal, state, or local in nature. And, there is a hierarchy depending on which agency takes the lead in the approval process; certain ordinances and regulations may be overruled or superseded by a higher authority. Political considerations can be a little more difficult to deal with in that they are mostly unwritten. They can include adhering to unwritten, but accepted policies, local customs, quid pro quos, and "doing the right thing."

At the outset of a project, the routing engineer should try to determine who the lead agency will be, the likely contributors to the certification process, and the probable timing for any approval. For some states, approval can take over 18 months. A close working relationship should be developed and maintained with the many agencies, organizations, and individuals involved with the project. One important factor to be aware of when dealing with agencies that must enforce various regulations, is the wide disparity that exists in interpreting those regulations. Many regulations are written ambiguously, intentionally or otherwise, with many gray areas. A US Forest Service (USFS), Bureau of Land Management (BLM), or county official in one area of the country may see things differently from his or her counterpart in another area, and not just because of the responsibilities for the different geography or natural resources.

It is a fact of life that each individual will have his or her own biases, prejudices, interpretations, responsibilities, and budgetary limitations. This must be understood and taken into account. An agency could have had a bad experience with the same company, or perhaps a different company, the last time a pipeline or other linear facility was constructed in its domain. The line may not have ended up where agreed, cover over the line may not have been adequate, or construction damages and mitigation measures may not have been carried out as promised. This, of course, could also happen with private landowners, but the outcome of lack of cooperation and respect can still be the same. Therefore, it is usually a good idea at the outset for the routing engineer to find out about past projects in the study area, how they were perceived by those affected, what poor images still exist, or what good things were done. A review of other EISs or EIRs is a good start.

As a routing engineer, however, one must be very careful not to make any deals when selecting a particular route that could backfire or jeopardize the project and one's professional integrity during regulatory hearings or in court. This could include changing the pipeline route from one unwilling agency or owner's property to an unwilling adjacent owner for no better reason than pressure. The best route must be defensible, adhering to the axiom "most public good, least private harm." The NIMBY syndrome (not in my back yard) is almost universal, but is not of itself sufficient reason to route a pipeline somewhere else. This is not to say that a political decision, at a higher level than the routing engineer, will not be made later to move the line elsewhere. The routing team can only go so far, making the best decision with the best available information.

In the section on the public involvement program process, we identify other intervenors, apart from NIMBYs, and see how and why they try to oppose a project. However, rather than just looking at who may derail a project, the pipeline proponents should concentrate on determining and cultivating supporters of the project. They can often be the greatest, objective, and independent allies, who may be in a better position to convince— or at least develop a consensus with—the detractors.

AGENCIES

Here is a partial list of agencies that could be involved in a pipeline project. The level of involvement from these various agencies might depend on whether the pipeline is interstate or intrastate, a replacement project, the length and size of pipe, or whether the proposed location is on private property, public lands, or in franchise.

Federal Agencies

Bureau of Indian Affairs—BIA
Bureau of Land Management—BLM
Department of Transportation—DOT
Environmental Protection Agency—EPA
Federal Emergency Management Agency—FEMA
Federal Energy Regulatory Commission—FERC
National Park Service—NPS
Office of Safety and Health Administration—OSHA
United States Bureau of Reclamation—USBR
United States Corps of Engineers—USCOE
United States Fish and Wildlife Service—USFWS
United States Forest Service—USFS

State Agencies

Coastal Commission
Department of Transportation—DOT
Department of Water Resources—DWR
Public Utility Commission—PUC
State Fish and Game—F&G
State Historic Preservation Office—SHPO
State Lands Commission—SLC
State OSHA
State Parks and Recreation

Local Agencies

Agriculture Board
City Council
Conservation District
County Board of Supervisors
County and City Planning Department

Fire Marshal
Flood Control and Irrigation District
Reclamation Board
Utility District

Local Political

Large landowners
Senators
House Representatives
Supervisors
Mayors

National or Regional Special Interest Groups

League of Women Voters
Sierra Club
Wildlife Federation

Religious or Tribal Interest Groups

Varies by regional location

Local Special Interest Groups

Business Round Tables—Kiwanis, Lions, etc.
Chamber of Commerce
Environmental
Historical
League of Women Voters

THE PUBLIC INVOLVEMENT PROGRAM (PIP)

Deciding on the best route and alignment may often be the easy part. The difficult part is convincing the agencies, public, and landowners of your selection; obtaining any sort of consensus will be an uphill battle if the project team has not worked with them throughout the process. Experience has shown that opponents of linear projects generally outnumber advocates by anywhere from 3:1 to 20:1.

It is very important at the outset of any project to involve the public—a community outreach—to get their input, to listen, and to gain productive communication. Giving out information to them is not the main intent. One should not confuse this public interaction with that which is required during the approval process when agencies and the public are asked to comment formally on environmental assessments. The PIP is a voluntary process instigated by the pipeline proponent during the initial phases of the project. Any public meetings should be handled delicately to avoid them turning into antipipeline rallies (see Format of Meetings and Holding Meetings later in this chapter).

Modern-day communications and access to a wealth of data allow the public to be extremely well informed about any "goings-on" in their area. Many are well educated and tend to get actively involved. Owners of property that may be crossed by the line can be civil engineers, attorneys, city planners, architects, biologists, or gas engineers. They often know, or think they know, as much as the project proponents about where a pipeline should be routed, or if it's needed. Many certainly know all the steps required to get a project accepted, and their legal rights. They must be listened to and respected. Be aware that active listening and public input do not necessarily mean compromising, giving in, or agreeing. The idea is to create a "Win–Win" situation, rather than "Win–Lose" where both sides end up in court and cause costly delays. Provided that the need for the pipeline has been established, the pipeline proponent has a duty to provide a needed commodity for the benefit of the public at large, and is usually in the best position to see the big picture. Again, the tenet "most public good, least private harm" must prevail.

When dealing with the selection of feasible routes in Chapter 2, the three Es—Engineering, Environmental, and Economic—were considered. In dealing with the public, there are another three Es—Equity, Empathy, and Effort. Equity means acknowledging the public's concerns, and legitimizing what they say about the project. Empathy is showing a genuine concern for their feelings about the project. Effort is a commitment to do what you promised to do, perhaps to follow up on a request for information.

Much opposition to projects, during the initial phases, is based on misinformation, but that does not mean that the answer is to bombard opponents with a mass of information. Some of the most voiced concerns and critical areas of misinformation are:

- Need—this project isn't needed; you already have sufficient capacity!
- Property values—my house, or land, will go down in value!
- Cost—we'll have to pay somewhere for this project!
- Decision on route—you've already made your mind up where the pipeline's going!
- Route alignment—I hear it's going right through my house!
- Trust—we heard that last time and you didn't keep your promise!
- Safety—these pipelines explode and are dangerous!

As with all aspects of life there will be detractors to the project, those people who make it their job to make yours as difficult as possible. The good routing engineer will learn to recognize them early on and how to handle them when asking for input, making a decision on the route, or explaining the decision.

As mentioned before, opponents of a project usually outnumber supporters by a wide margin, but are not necessarily all active in their opposition. Often, supporters are strong proponents of change and see new projects as generating more jobs and improving the status quo. Detractors may have several different agendas:

- NIMBY (not in my back yard)—not necessarily against the project; they just don't want it on their property.
- Environmental—quality of life must not be degraded. Support may be offered if project proponents provide considerable mitigation.
- Slow growth/no growth—concern that pipeline will induce unwanted growth. Opponents believe that "green" energy or conservation is what's needed.
- Land use—a pipeline is not acceptable as a suitable use of the land.

Hoping that certain opponents will see the clear benefit of the project, if explained to them rationally with detailed advantages, is usually totally unrealistic. Opponents must be acknowledged and handled appropriately. To think one can make an end run around them by ignoring them and not confronting their concerns is to jeopardize the schedule, and perhaps the project.

For those who have been involved in the pipeline industry for many years, and especially in the less regulated or new growth states, the preceding may seem somewhat far-fetched. However, knowledge and understanding of the people whose lands you wish to cross are every bit as important as information about the topography, property boundaries, and construction details.

The stage at which a public involvement program is initiated can be critical to the success and/or timing of project completion. Each project could be different. Involving the public too soon could stir up unnecessary concerns from too many people who may never be affected by the final alignment. Do you select the terminal points and then go to the public with a blank map and ask for suggestions on how to get between them? Probably not. The public may be well educated, intelligent, and knowledgeable about parts of the study area, but few will be privy to the vast amount of information, including constraints, that can and should be gleaned by the pipeline proponents. The blank map approach is to invite a million and one different routes from many different people, all with different ideas as to where a pipeline should or shouldn't go. A study area should first be defined that is reasonable and acceptable to the pipeline company, yet not so restrictive that it would exclude interesting solutions, which may include paralleling existing linear facilities. At least two or three corridors should then be selected, as outlined in Chapter 2, before going public. This approach will provide focus to the first meeting with the public, who will undoubtedly propose additional alternative routes for study. Some may have merit; others may not.

Setting up a public involvement program can be as comprehensive as you wish and take many forms. Basically the PIP steps are:

1. format of meetings and selection of venues;
2. public notification;
3. holding meetings; and
4. communication and follow-up.

Format of Meetings

How a meeting is held may be dependent on type and size of the project, the area and kind of people you are working with, and the project participants' level of comfort. Meetings can also be a combination of various types. These include: one-on-one meetings in someone's office, or individual door-to-door; coffee klatches with neighborhood groups; informal, open-house meetings at a community center or school; or, the more formal, town-hall type meeting. An advantage of the open-house approach is that it allows a more informal interaction between the project proponents and the public. Shyer members of the public are much more inclined to ask questions and express their concerns; of course, questions can still be destructive as well as constructive, and suggestions offered both subjective and objective. In the formalized town-hall meeting there is greater polarization between proponents and opponents, where a few strong characters tend to monopolize the proceedings and posture in front of the rest of the public.

Choice of meeting location is important. You want maximum exposure to the public, so choose a place that is easy to get to, and large enough to hold the anticipated audience. Trying to keep low key, hoping only a few people turn up, will only cause problems later on in the process as frustration of not being "in on things" ultimately leads to obstructionism.

Public Notification

Notification of meetings may be done in a variety of ways. A project mailing list must first be developed, which includes the names and addresses of lead and cooperating agencies, potentially affected public, federal, state, and local agencies, and environmental organizations (see Chapter 5 for examples). As mentioned previously, one way of notification is through the Federal Register. For those on the mailing list, an initial newsletter may be circulated. To cover other interested parties, or those you may have missed, consider using advertisements in several local newspapers, time slots on local radio or TV, or placing notices in prominent places throughout the study area, such as community centers, large local stores, and gathering places.

Be aware that no matter how much notification you think you have given the public, it never seems to be enough. This is one of the hottest issues when receiving comments from the public. Attorneys for landowners or intervenors will often claim lack of due process saying either they never knew about the project, or were given insufficient notice.

Holding Meetings

As with the venue for a meeting, selection of days of the week and time of day to hold meetings can be an issue in ensuring maximum public participation. The format will differ between the more formal town-hall gathering approach and the more informal open-house approach. For the former, each meeting usually begins with introductions and presentations by members of the project team. They will address project description, purpose and need for the pipeline, the environmental planning process, the study area and preliminary alternative routes, and the project schedule. The floor is then opened up for comments and questions. There will be many questions that can be answered simply, but there will be a number, including combative statements and comments, that should be left alone. With these it is best to listen, noting the gist of the remarks on a flip-chart. Mounted maps and aerial photographs of the study area, showing the routes, and other photographs or diagrams showing construction details, may also be displayed in strategic positions.

The open-house format allows the public to visit the meeting place, at a time that suits them, over a period of say three or four hours. The project team experts are available throughout at booths or tables for one-on-one discussions on the subject matter that members of the public are interested in. For example, there could be various booths that cover the routes, environmental issues, right-of-way, and property appraisal questions, or construction details. The public then has the opportunity to have their specific concerns addressed. Both formats should provide a method for them to give their names, addresses, and telephone numbers so that they can be apprised of the project's progress.

These informational gathering meetings can be invaluable to the routing engineer and team members who are willing to listen. They can then sift through what has been heard, compare notes with the project team's information and thoughts on routes, and determine what changes, if any, should be made. The routing engineer should always remember that there are members of the public willing to give constructive and objective information; many may be long-time residents in the study area and have knowledge and insight that the project team could not possibly have. Use these meetings as a very positive experience to listen and learn, rather than a process to be tolerated and "got through" as soon and painlessly as possible. These will not necessarily be comfortable meetings, and often one's credibility, expertise, and composure will be challenged.

Communications

The initial communication with the agencies and public may be the announcement of the first project information meeting. Thereafter, there must be a variety of follow-up materials or methods. Newsletters, fact sheets, and project updates are common, and the project manager may wish to set up an 800-number hotline for supplying up-to-date information and answering questions. A good faith effort to respond to all questions goes a long way to gaining the public's confidence and acceptance. Offering to meet the public and/or agencies at a location of their choice cannot be underestimated in establishing a rapport for the exchange of useful information and ideas.

Whatever format and method of communication is used, all questions, answers, and comments should be prepared soon after meetings. This provides documentation of the PIP process.

Chapter 5
ENVIRONMENTAL CONSIDERATIONS

INTRODUCTION

During and immediately after World War II pipelines were relatively short and pipe sizes small. Over the years, technological advances in the manufacture and construction of pipe, and a recognition of the economic advantages of pipeline transportation, have changed that situation to one where numerous products can now be carried long distances. With increases in pipeline length and size, the public became increasingly aware of the environmental impact such lines might cause. Today, pipeline companies face the task of locating and constructing lines to meet increasingly complex environmental protection guidelines. This places an increased responsibility on all project team members, especially the environmental planner, to provide essential supporting data to the routing engineer.

Natural resources' (natural gas, water, petroleum products, etc.) production, transmission, and storage systems play a critical role in the nation's economic and social well-being. Many utility and/or pipeline company customers take their operation for granted as they enjoy natural product services relatively free of interruption. There is an increasing need in the United States to work cooperatively to maintain greater resource and transmission flexibility and to enhance service reliability through transmission system interconnections.

Utility companies are responsible for supplying adequate supplies of reliable economic resources to their customers. Although gas, water, and oil can be stored in a number of different locations, gas and oil especially are discovered and developed in regions that are seldom close to where load requirements exist. Therefore there is a considerable need for transmission lines, often quite long, to provide for interregional transfers of these resources.

PURPOSE AND NEED

To help in following the rest of this chapter two flowcharts, Figures 5-1 and 5-2, show a typical pipeline study process.

Federal regulatory agencies have discretionary authority over the sale of certain natural products, such as gas, oil, and water, and the selection and design of new or upgraded transmission facilities. Their review considers the need for the products, the pricing rate of their sales and transmission systems, and the consequences of new transmission systems and corridors.

As mentioned in the introduction, the goal of utility companies is to provide reliable services of their products at the lowest reasonable infrastructure cost. On the other hand, the state and federal regulatory agencies' responsibilities are to establish rules and review the proposed actions of utility companies to ensure that consumers are provided service at the lowest possible costs. Recent industry initiatives to minimize costs have focused on three areas:

- integrated least-cost planning;
- free enterprise in a deregulated competitive market; and
- environmental costs.

Transmission lines must meet two tests to be shown beneficial to society: environmental impacts and consumer benefits. Until a project has cleared environmental hurdles it is not considered prudent to include it in least-cost plan alternatives. Utilities cannot make plans to meet service requirements without some confidence that a resource option will be possible. Furthermore, to do so would presume a favorable decision through the National Environmental Policy Act (NEPA) process. As environmental costs become an important consideration in the resource planning process, low environmental cost resources become more important. Remember that many environmental impacts can be offset or translated into monetary costs.

Regarding the consumer test, utilities must show their regulators that a new transmission line will benefit the public at least cost. Note that total costs need not necessarily be reduced; witness the replacement of a failing pipeline or addition of capacity due to increased demand. In addition, the purpose and need for a new pipeline must be demonstrated to the regulatory agencies and the public; if not, any selection of corridors, let alone routes, will be a waste of time.

SCOPING AND PROJECT-RELATED STUDIES

Although not all pipelines will come under the jurisdiction of the Federal Energy Regulatory Commission (FERC) or state public utilities commission (PUC), we will assume so in order to make this manual comprehensive and

ENVIRONMENTAL CONSIDERATIONS 59

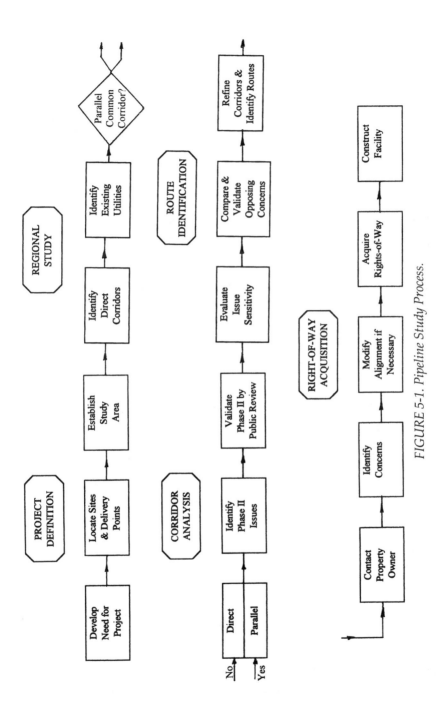

FIGURE 5-1. *Pipeline Study Process.*

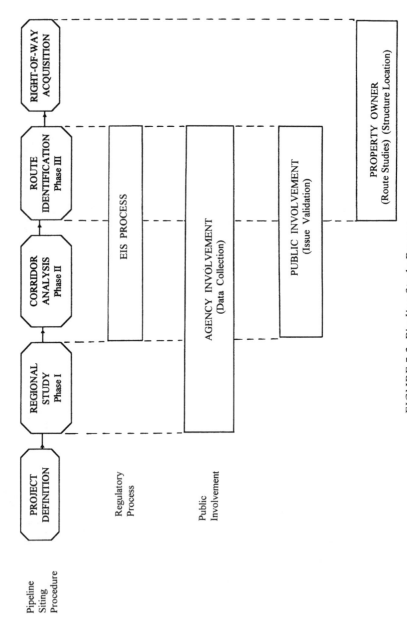

FIGURE 5-2. Pipeline Study Process.

include all possible scenarios. A federal project requires an Environmental Impact Statement (EIS) and a state or local project (depending on the state involved) requires an Environmental Assessment (EA) or Environmental Impact Report (EIR). A Draft EIS or EIR is first prepared, followed by a Final EIS or EIR.

As required by NEPA (1969), and the Council of Environmental Quality (CEQ) regulations (1978) for implementing NEPA, an extensive coordination program should be developed to ensure that all the appropriate members of the public and federal, state, and local agencies are contacted, consulted, and given an adequate opportunity to be involved in the process (Chapter 4 described the public information and participation programs). NEPA also requires that "reasonable and feasible" alternatives be compared during an EIS.

Federal departments and agencies, such as the U.S. Department of the Interior (DOI), Bureau of Land Management (BLM), Department of Agriculture (USDA), Forest Service (USFS), DOI Bureau of Reclamation (BOR), Bureau of Indian Affairs (BIA), and National Park Service (NPS) must complete numerous scoping activities before a certificate can be issued for construction of an *interstate* pipeline. Scoping is an information-gathering process open to the public early in a project to identify the range or scope of issues to address in the ensuing environmental studies. Scoping serves to identify significant issues to be analyzed, determine how they will be treated in a DEIS, and to eliminate issues and alternatives from detailed study, where appropriate. Scoping information from agencies and the public provides the basis for identifying alternative routes and developing the work plan for environmental baseline, impact assessment, and mitigation planning for the project.

Scoping activities include:

- review of previous studies of transmission projects in the area;
- completion of a regional siting study, including resource sensitivity analyses, agency contacts, and public scoping meetings;
- identification of project issues; and
- identification of alternative transmission line corridors and routes.

If the lead agency determines that there will be impacts, and an EA is insufficient, a Notice of Intent (NOI) to prepare a DEIS for a pipeline project will usually be published in the Federal Register, and public scoping meetings will be held soon after in selected communities throughout the study area. The lead agency will handle this process.

Intrastate pipelines may have to comply with a similar process in certain states subject to the regulations of a PUC. For example, regulated utilities in California must conform to CEQA guidelines—the California Environmental Quality Act, enacted in 1970.

ENVIRONMENTAL COMPLIANCE OR CONSIDERATIONS

Today, the environmental compliance process is a fairly straightforward affair, inasmuch as the guidelines for an EIS or EIR are more or less standardized with respect to format and what must be addressed. Unfortunately, interpretation of certain elements, and adherence to the set schedule, can be widely abused in some areas, due partly to inexperience, overzealous officials, or hidden agendas. However, in addition to addressing the purpose and need for the project, description of the project and alternatives (which includes alternative corridors and routes), and design, construction, and maintenance, an assessment must consider environmental impacts on these major resources.

Natural Environment:

- plant habitat,
- soils and geology,
- surface hydrology and wetlands,
- threatened and endangered plant and animal species, and
- wildlife habitat and use areas.

Human Environment:

- existing, planned, and designated land uses;
- parks, recreation, and preservation uses; and
- scenic and aesthetic resources.

Cultural Environment:

- archaeology,
- ethnohistory,
- history, and
- prehistory.

A typical outline for the contents of an environmental assessment, which includes these broad categories, might be:

- air quality,
- biological environment,
- cultural environment,
- cumulative impacts (considering other projects in the area),
- geology, topography, soils, and seismicity,
- growth-inducing impacts,
- hydrology and water quality,
- land use,

ENVIRONMENTAL CONSIDERATIONS   63

- noise,
- public health and safety,
- public services and utilities,
- socioeconomic environment,
- socioeconomics,
- transportation, and
- visual resources.

A sensitivity analysis is then completed, and opportunities and constraints are listed or mapped for potential alternative routes. Sensitivity is a measure of the probable adverse response of each resource to direct and indirect impacts associated with the construction, operation, maintenance, and abandonment of the proposed pipeline.

The preceding list is wide in range, requiring the input of many experts from different disciplines. Although not expected to be an expert in any one of these, the routing engineer, who wants the greatest chance of success in selecting an acceptable route, should know enough about these disciplines to ask the right probing questions. The routing engineer should have an innate ability to see the big picture, and coordinate and balance the experts' information as objectively as possible. A route analysis process and route matrix comparison (again see Table 2-2) can be carried out to help the team select a recommended route. Consider this an independent and somewhat mathematical solution, especially if weights and numbers are assigned to the various criteria.

CORRIDOR STUDIES

Alternative pipeline transmission corridors are identified based on previous studies, a regional siting study, and public and agency input. Subsequently, the environment is inventoried and the data compiled in numerous layers along all final alternative corridors. Increasingly, for many projects, these layers of information are being placed within a Geographic Information System (GIS). These baseline data are then used for assessing project-related impacts. When initially studying corridors a check should be made to see if there are any designated utility corridors in the study area that are going in the direction you want to go. The Federal Land Policy and Management Act (FLPMA) of 1976 mandates, to the extent practical, that BLM will consolidate future utility projects within a corridor that has been established already. The BLM or lead agency's difficult mission is to protect the quality of the land resources, environment, and public values while permitting development and use of a linear facility, such as an underground transmission line, in a cost-effective manner, and which will help meet society's needs.

After this stage, public workshops are held to report results of the environmental studies, present preliminary alternatives, and gain public input regarding the acceptability of those alternatives.

ROUTING ALTERNATIVES

Final routing alternatives for the proposed pipeline are determined through a process of documentation and elimination of alternatives with serious constraints. Alternative routes are eliminated for a number of reasons, including environmental conflicts, public and agency opposition, and system planning and performance criteria.

To select routing preferences, the environmental consequences of each route are summarized, based on impact assessment results, environmental resource preferences, and agency and public comments.

ENVIRONMENTAL CONSEQUENCES

The consequences to, or impacts on, the environment caused by implementing the proposed project are assessed by considering the existing condition of the environment and the effects of the activities of the proposed project (construction, operation, and maintenance) on the environment. The initial impacts are evaluated to determine if mitigation measures will be effective in lessening the impacts. Those impacts remaining, after applying mitigation measures, are referred to as residual impacts. Some impacts are considered adverse, direct, and long-term. Other impacts (e.g., visual, some cultural, and biological) are considered adverse, indirect, and long-term.

One should note that environmental documents or reports usually also describe a "No Action" alternative. This outlines the potential and diverse impacts of not building the project.

Earth Resources

The principal impact associated with earth resources is the potential for increased erosion hazards, or soil and water contamination from leaks or spills. Some short-term soil compaction impacts can occur in agricultural areas, and stream sedimentation can also occur at crossings of perennial streams.

Biological Resources

Typical impacts to biological resources include effects on threatened, endangered, or protected species, rare or unique vegetation types, migration corridors for wildlife, areas of low revegetation potential, or highly productive wildlife habitat. The impacts are generally associated with removal of vegetation and habitat caused by construction and operation activities, and from human activity that allows more access into remote areas.

Land Use Impacts

These impacts are those that would displace, alter, or physically affect any existing or planned residential, commercial, or industrial use or activity, any agricultural use, or any recreational, preservation, educational, or scientific facility or use.

Socioeconomic Effects

These include construction period impacts to area communities, social and economic impacts along the selected route, and fiscal impacts with local jurisdictions. These effects can be both adverse and beneficial.

Visual Impacts

Considered adverse, direct, and long-term, visual impacts include effects to the quality of any scenic resource, the view from any residential or other sensitive land use or travel route, or the view from any recreation, preservation, education, or scientific facility. Most visual impacts for a pipeline will be short-term during the construction phase, except, of course, ancillary facilities, such as compressor stations. Unlike an electric power line, where tall structures will be in evidence after construction, in most instances a pipeline will be buried with little evidence of its existence after a fairly short time and with proper revegetative mitigation. However, there are occasions, depending on type of vegetation, soil, and geology, where the construction scar may remain for 20 years or more. Adverse visual effects may occur at sites with high aesthetic or interpretive values.

Physical Impacts

Direct, adverse physical impacts can occur to cultural resources during construction, whereas indirect impacts can result after construction due to increased erosion or increased access to sites.

PREFERRED ROUTE SELECTION

Based upon review of potential impact characterizations, significant, unavoidable adverse effects, agency and public comments, and cumulative environmental consequences of the alternative routes, the preferred route(s) is then identified. Always remember that many adverse impacts can be mitigated to less than adverse, and that a particular route should not be eliminated out of hand.

Chapter 6
ACQUISITION OF LAND RIGHTS

SECURING LAND RIGHTS

The first contact that a right-of-way agent is likely to make with a landowner is gaining permission to enter the property to carry out preliminary land studies. This usually covers surveyors staking a preliminary centerline so that biologists and cultural resource specialists can gather information for the EIS or EA. There have been several recent cases, in various parts of the US, where landowners have been heavily fined for contravening federal statutes on wetlands and endangered species. This has resulted in a growing reluctance on their part to allow environmentalists on their land to perform studies. They are concerned that a rare plant or endangered species will be discovered that prevents them doing with their land what they feel is their right. In some cases, private landowners have asked project proponents to sign indemnity agreements, stating that if a rare species is found, and they are unable to develop their property, compensation will be paid—naturally an unrealistic request.

In other cases, the routing engineer and/or surveyors, usually the first people on the ground, have been required to show evidence that their qualifications do not include an environmentally related degree before they will be allowed on the property. Where the potential for hazardous waste conditions exists, property owners have allowed entry for preliminary surveys, and even biological studies, providing no core sampling or related activities are performed. Such situations are important to know about, so that entry agreements, with specific legal language acceptable to both parties, can be tailored to allow the earliest entry to properties with minimum delays.

Once a preliminary and suitable route (from an engineering and environmental standpoint) has been chosen, the routing engineer should have the property records reviewed to determine existing easements. These may not be apparent from walking the route, but their existence, especially if exclusive, have excluded many a route from further consideration.

After the alignment of the pipeline and its ancillary facilities have been selected, land rights must be secured before any construction activity begins. Land rights include easements or fee property purchase from private landowners, and permits or licenses from those companies or governmental agencies who can only grant these types of land rights.

It is highly recommended that land rights are acquired which are specific to the alignment of the pipeline, and the restrictions on use. Where required land rights are of such nature that they make lands outside the take severely unusable, or minimize the value to the fee owner, additional takes may be required or damage payments made.

The documents that are prepared should describe the land over which the rights are to be acquired, the owners of the land (which may include lessees, minerals rights owners, and lien holders), a legal description of the land rights to be acquired, and the restrictions in the strip or area of the land encumbered by the pipeline. The land rights should also contain language allowing the pipeline operator to grant the rights to successors of the pipeline owners, and include rights of ingress and egress for maintenance and operations. They may be blanket or specific rights.

The descriptions should be based on a land survey of the pipeline alignment, and a width of the perpetual rights to be acquired as well as a description of temporary construction rights. Often the centerline of the proposed pipeline serves as the description of the right-of-way centerline, but the described line can also be an offset line to allow for future parallel pipelines.

The land rights to be acquired may vary from full fee over the land, exclusive rights (which will require other future land rights to be granted over the strip subject to acceptance of the pipeline operator), to standard perpetual or time-constrained easements.

The land rights documents should be prepared in such a form that they are legally binding on both parties, and should be recorded in the same place and manner as deeds for property sales. Public notice of the documents is important, because pipelines are not visible from an inspection of the land, except for intermittent markers; therefore they serve no constructive notice of their existence. This is also important if there is a delay between acquisition and construction. It is good practice to attach a map to the legal description, showing the relationship of the pipe to certain features. This could be simply diagrammatic, not to scale, or an as-built scaled drawing. A legal description may mean something to a surveyor or lawyer, but is of little value to the majority of landowners for understanding where the pipeline is located. A digital orthophoto, or rectified aerial map, overlaid with the pipeline centerline, bearings and distances, right-of-way dimensions, and property boundary lines, can be of great value.

Payment for the land rights to be granted should be made after an appraisal of the full fee value of the lands crossed; the present and future use is taken into account and the value expressed in dollars per acre. This dollar figure serves as a basis for the percentage of compensation for the rights to be acquired. Annual rental fees should be avoided whenever possible due to the uncertainty of escalation of the fees. This may be a factor in routing a pipeline away from entities, such as railroads, that often require an annual fee; unless, of course, current and future economics are favorably negotiated. Special compensation can be made on crop losses and damages to land due to construction activities. Construction damages fall into two categories: short-term and long-term. Short-term are temporary in nature; for example, removal of a row of vines in a temporary working strip, which will be replaced later after construction is complete. Long-term would be the permanent removal of vines within the right-of-way. Note that in certain parts of the country it is still common to make right-of-way purchases on a dollar per rod basis. This tends to avoid the premise that the property is permanently damaged.

If the acquiring agents make special agreements with individual property owners, these should be made subject to the approval of the pipeline operator. Special agreements may not be spelled out in the recorded document, but they should be in writing with both parties retaining copies.

The acquiring agents must be fully informed of proposed construction activities and engineering criteria so that they can present and discuss issues with property owners, their representatives, or lessees. At the outset of negotiations, a large scale aerial photograph covering the property owner's land can be of inestimable value in pointing out where the pipeline is intended to be laid. Experience has shown that leaving a photograph with the owner is good public relations at minimal cost.

An acquisition agent's demeanor is also very important. Because the agent often serves as the first contact with the property owner, any prolongation of acquisition time will affect a project's schedule.

USE OF FRANCHISES

The indeterminate franchise ordinances conferring the right to occupy streets, roads, and ways are passed by local jurisdictions, such as cities and counties, and contain express conditions that the pipeline or utility company will relocate its facilities that interfere with "…lawful changes in grade, width, and alignment…" of the roads occupied. This relocation requirement is also interpreted to include ancillary facilities of roadways such as storm drainage devices. In most cases, careful engineering study of possible future improvement of occupied roadways, and allowance for these improvements, will allow pipelines to remain undisturbed in the road right-of-way for their functional life. The rights conferred by franchise ordinances are nonexclusive, and other facilities may occupy the general area occupied by the pipeline. However, safety regulations as to spacing and the like apply, so conflicts and operating inconvenience should be minimal.

There is usually no direct cost for installing a pipeline in a franchise area since the various methods used by pipeline or utility companies for computing payments, as provided in the various franchise ordinances and state statutes, are based on revenue generation, either by the company as a whole or within the local jurisdiction of the franchisor. Use of franchise areas for pipeline installation is generally more expensive from a construction standpoint since the work takes place in, or very near, an improved roadway in use by the public. Use of franchise areas may be desirable through areas where surrounding private lands are subject to flooding, are ripe for urban development, or are very expensive.

Careful research, and a legal opinion, may be needed for roads that appear to be private. Dedications, often made many years ago, may have rendered them as franchise areas. In a similar vein, do not automatically assume that railroad companies own the underlying fee to the strip of land where their tracks are laid and can prevent a parallel or perpendicular pipeline placement.

PROPERTY OWNER RELATIONS

Pipeline easements across private land restrict some land uses within the right-of-way. The property owner is normally compensated for these land use limitations at fair market value considering the highest and best use of the property. Pipeline rights-of-way generally prohibit underlying fee owners from constructing structures, drilling wells, removing or adding ground cover, and planting deep rooting vegetation. Most other uses are permitted as previously discussed. From the point of initial contact with the property owner for permission to enter to perform engineering and environmental surveys, through the right-of-way negotiation process, the property owner has an opportunity to discuss compatibility issues. Conflicts with agricultural planting, harvesting, range management, irrigation systems, road networks, and other concerns should be investigated and an effort made to resolve them. In addition, the property owner is normally reimbursed for substantiated claims of incidental damage resulting from pipeline construction and operation activities.

Often, the right-of-way agent's and the owner's idea of fair payment for the property to be taken is poles apart, with little chance of an agreeable settlement. Provided that a good faith effort has been made by the pipeline company, over a reasonable period of time, to offer fair market value based on a professional appraisal, but has been rejected, the pipeline company has two options. If it is a public utility or common carrier with powers of eminent domain, it can go to court to condemn for rights and an order of possession for construction. Without these powers, it either has to pay the going rate, or route the pipeline across the land of a more amenable owner. In some cases the latter may not increase line length; if it does, there will be a commensurate increase in pipe and construction costs. Truly a seller's market!

However, the power of eminent domain cannot be exercised over Federal Government lands; this includes those owned by Native American tribes. As Nation States, they are independent and not subject to most federal and state laws. Straight payments are often not the only factor in acquiring land rights, and many companies have found it cheaper to go around reservations, the exorbitant amounts asked bearing no relationship to what might be perceived as fair market value.

In the same way that the public has become much more sophisticated in understanding the corridor selection process and planning issues, so have landowners in maximizing revenues from an easement grant and related damages. Many simply turn all negotiating over to their attorneys, who usually recommend going to court to elicit the highest settlement price. In addition, payments for an easement are taxable, whereas damages are not.

For an additional treatise on this subject please refer to ASCE Manual 75, *Right-Of-Way Surveying*.

Chapter 7

CONSTRUCTION, MAINTENANCE, AND OPERATION

TYPICAL CONSTRUCTION ACTIVITIES

The pipeline routing engineer should have an understanding of the typical pipeline construction sequence for the type of pipeline to be constructed. As an illustration, a typical steel cross-country pipeline spread might proceed with the following steps (see Figure 7-1).

Mobilization

The pipeline constructor moves in labor and equipment and establishes a field office and laydown yard for equipment and materials.

Clearing of Work Strip

A work strip for the pipeline right-of-way is normally cleared to provide room for construction equipment. The clearing work is performed in accordance with permits and agreements made with each landowner or permitting agency. Using an onground marked line, the construction work strip is cleared and graded. Vegetation and obstacles are cleared to the extent necessary to allow safe and efficient use of construction equipment (see Figures 7-2 and 7-3).

Clearing of the work strip should follow accepted industry practices and sound construction guidelines. In areas where timber clearing is required, the trees should be cut to uniform lengths and stacked along the right-of-way based on the owner's preferences. Stump profiles should be kept as low as possible. Stumps should be removed, as needed, before pipeline installation. Debris created from right-of-way preparation should be disposed of using approved methods.

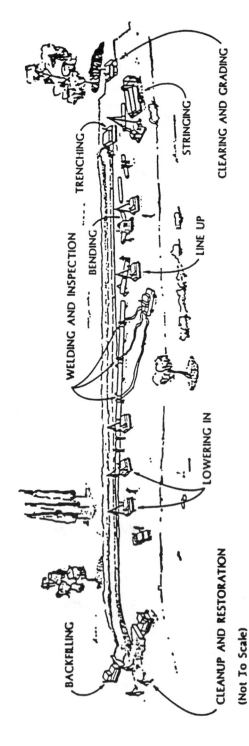

FIGURE 7-1. *Typical Pipeline Construction Sequence.*

CONSTRUCTION, MAINTENANCE, AND OPERATION 75

FIGURE 7-2. Flagging C/L Final Alignment and Edge of Construction Work Strip.

FIGURE 7-3. Cleared Swath Through Corn for Construction Right-of-Way.

Where irregular terrain is encountered, a work strip, wider than that required for level terrain, is needed so that pipeline equipment can work off a flat area perpendicular to the pipeline centerline (refer back to Figure 2-10).

Trenching

The typical trench provides the minimum required cover and side clearance. Deeper trenching may be needed at selected locations (such as stream and road crossings) to provide additional cover. Trenching generally is accomplished by wheel trenchers and backhoes. Where hard rock is encountered, it is either blasted and removed by backhoe, or cut by a special trencher with rock-cutting teeth. Spoil removed from the trench is normally left alongside the trench in a spoil bank. A safe location for the spoil bank, and shoring requirements for certain depths of trench, may be subject to various state safety regulations. Where the pipeline is in a paved road, the spoil is usually hauled away to an approved landfill site. For rocky areas, approximately 150 millimeters (6 inches) of fine rock-free soil is normally spread in the bottom of the trench to pad the pipeline.

Spoil Disposal

The nonhazardous wastes generated during pipeline construction could include soil, asphalt, rocks, concrete, sanitary wastes, and tree stumps or branches. Excess soil or rocks would normally be spread over disturbed areas along the pipeline route. Any remaining nonhazardous wastes should be disposed of using approved methods, or removed to a Class III landfill site. Sanitary wastes should be removed from the construction site and disposed of in appropriate septic facilities.

Without establishing an exact alignment, it is not possible to estimate the quantities of nonhazardous materials that may have to be disposed of at landfill sites. However, with consideration of routing the pipeline in open spaces, and by avoiding rock outcrops, only small amounts of material need be disposed of as a result of the pipeline construction.

Installing the Pipe

Typically, 12 meter (40 feet) long pipe segments are strung on wooden skids alongside the trench. The pipe is then cold bent to conform to the trench profile; the joints are bonded, and then welded together. This is followed by radiographic inspection of the welds to ensure weld quality. The entire pipeline coating is inspected for uniformity and the welded pipe string lowered into the trench. Typical equipment used to install the pipe consists of pipe trucks, welding trucks, x-ray trucks, pipe bending machines, pickup trucks, and side-boom tractors (see Figure 7-4).

FIGURE 7-4. Sidebooms lowering 500 mm Pipeline into Trench.

Backfilling

The initial backfill—300 millimeters (12 inches) over the pipe—should be composed of clean fine native soil or select clean imported sand or soil. Where native soil contains rocks, a shaker is used to produce the fine native initial backfill. After the initial backfill is placed, the remaining trench is filled with native soil. Excessive amounts of rock are hauled away to an approved landfill site. Smaller rocks are spread evenly across the work strip.

Strength Testing

Typically the pipeline is pressurized to a minimum of 1.25 times its design pressure, for a minimum hold time of 8 hours, to ensure strength of the pipeline.

Cleanup and Work Strip Restoration

Cleanup and restoration activities can include recontouring the work strip and repairing fences, roads, drainage, and stream banks. To mitigate erosion, water breakers may be installed in sloped areas and disturbed areas seeded.

Even though this sample pipeline construction sequence is typical of a steel cross-country pipeline, others may require different joining, installing, inspection, and testing methods; however, the basic construction sequence is similar. There may need to be a number of modifications to this sample sequence for urban pipeline construction.

TYPICAL MAINTENANCE AND OPERATION ACTIVITIES

Periodic routine patrolling of pipelines is required. Generally pipelines are patrolled by airplane or helicopter to observe signs of leakage, construction activity, encroachments, or earth movement along the pipeline right-of-way that could potentially damage or create a hazard to the pipeline. The aerial patrol is usually supplemented by a ground patrol by vehicle and/or on foot.

Chapter 8
ECONOMICS

PIPELINE ECONOMIC CONSIDERATIONS

Pipeline systems are designed to provide the highest return on investment and lowest cost of service for an expected economic life. The economics for a pipeline system design are based on forecasts of costs and economic conditions as far out in the future as 40 years. The calculated optimum design for a pipeline system one year, under one set of economic conditions, may change significantly given a new set of circumstances.

The question, "To build or not to build" is the basic question that must be answered in any economic analysis of a proposed pipeline system. But, more important, is the question, "What to build?"

Selecting the pipe material, wall thickness and diameter, and other components of a pipeline system, is not just a matter of looking in a reference book for the line size corresponding to the desired throughput capacity. Persons responsible must visualize and describe the feasible options and then decide the best pipeline diameter, number and location of pump stations, types of prime movers, and other lesser but important variables.

The selection of the pipe diameter is obviously the most significant design consideration. The larger the line pipe, the fewer the stations and the simpler the operations. Conversely, the smaller the line pipe, the more numerous the stations and the greater the supporting requirements. Other features can be modified in time, if they are found later to not be the best choice, but changing the diameter of the line pipe and relocating stations for hydraulic balance after installation would require a complete pipeline system replacement.

There are many economic factors that influence and complicate the economic analysis of a pipeline system. Many publications discuss this subject in great detail. A good reference related to pipeline economics is the ASCE Pipeline Division Committee on Pipeline Planning Report, *Pressure Pipeline Design for Water and Wastewater: Section 7—Pipeline Economics*, 1992.

The pipeline routing engineer needs to have an understanding of the typical pipeline economic factors that have an impact on pipeline routing alternatives. Following is a listing and limited discussion of some of these economic factors and conditions.

PRIMARY ECONOMIC FACTORS

Primary economic factors are used as part of a comparative exercise to determine the most economically desirable design for a new pipeline system.

Estimated Cost of Building

Under the cost of building the pipeline system, the two most significant factors are:

- cost of line pipe, and
- cost of pipeline construction.

Together, these factors total about 75% of the cost of an average project.

The cost of building pump stations may or may not sway a decision on line size, but the features of a station are influenced by changing economics.

Estimated Annual Operating Expenses

The annual operating expenses are estimated for each year of assumed economic life. Many expenditures enter into the total annual operating expenses and costs of owning a pipeline system. However, only three have noticeable weight in calculations for deciding the diameter of a pipeline:

- cost of power,
- ad valorem taxes (property taxes), and
- cost of insurance

Power costs are calculated based on research of power sources available at each site. Ad valorem taxes and insurance are estimated as a percentage of capital costs. Comparative labor, and routine operating and maintenance costs between line sizes are not generally significant. Product loss, when considered an operating expense, would normally be assumed equal for any potential diameter pipeline.

Forecasted Escalation Rate

The escalation rate of the operating costs, and costs of building are estimated. The rate or pattern at which the annual costs are predicted to escalate each year is a significant consideration. Because the economic exercise used to evaluate design options usually involves discounting future costs to a comparative present value, the rate of escalation for the first few years of operation weighs heavier than changes in the later years of economic life.

Assumed Value of Money

The value of money is assumed as governing desired rates of return on capital and/or interest rates on borrowed capital. Capitalized financing costs can be a significant component of the total cost if a major system takes several years to build before the start of operations.

SECONDARY ECONOMIC FACTORS

Secondary economic factors are mostly subjective factors that influence if or how the primary economic factors are used in the alternative selection process.

Economic Weather

The economic "weather" is assessed at the time of decision. The economic climate may either make the owner bullish for a larger diameter system, or bearish for the smallest feasible diameter.

System Owner's Perception

The proposed pipeline system owner's changeable perception of the project is based on directly affecting changeable financial situations. The owner is subject to varying credit limits, alternative needs for money, and has a biased assessment of the risk of forecasted throughput and other related assumptions.

SUMMARY

The routing engineer needs to have a basic understanding of the economic factors that affect the evaluations and decisions related to whether a pipeline project gets installed. The list of significant pipeline economic factors can be summarized as:

- cost of main line pipe and pipeline construction;
- annual cost of operating power;

- annual cost of ad valorem taxes and insurance;
- forecasted rate of escalation of power, ad valorem, and insurance costs, particularly for the first 10 years;
- economic conditions at the time of decision; and
- the perception of the project owner and other current financial commitments.

Changes in these factors can change the diameter of the pipeline, number of stations, needed horsepower, and what other features will be installed.

A good reference that goes into more detail regarding this subject is an article in the *Pipeline & Gas Journal,* "Energy supply economics—changing economic factors affect pipeline design variables," by Wayne C. Wildenradt, July 1983.

GLOSSARY

AASHTO: American Association of State Highway and Transportation Officials.

Access Road: A designated path leading to the pipeline and/or related structures, accessed by maintenance crews. Any road to reach the work site.

Alignment: The actual location of a linear utility facility, such as a pipeline transmission line; the centerline of the right-of-way as established by a survey. *Note:* The centerline of the ROW will not necessarily be the centerline of the pipe, which may be installed closer to one edge of the ROW for maintenance purposes, or to make room for another pipe at a later date.

Angle Point: A point where the pipeline changes direction.

BLM: Bureau of Land Management. An office of the Department of the Interior.

Capable Corridor: A corridor that contains at least one potential pipeline route.

CEQA: California Environmental Quality Act. Passed in 1970 by state legislature, this legislation requires public agencies to consider possible environmental effects of a proposed project, alternatives to the proposal, and measures to mitigate any adverse effects.

Color IR: Color-infrared photography is useful in site selection. Vegetative cover can be determined, together with soil and moisture conditions, and seepage zones in landslide areas.

Constraint: Any existing or future physical feature or plan that either prevents or limits a particular route. Mitigation may be required if the constraint is not avoided. Examples might be a park, lake, or proposed subdivision.

Contributors: Those people involved in the development and execution of the public involvement plan for the project, including pipeline company representatives, and environmental and public involvement consultants.

Corridor: A strip of land of variable width, often one to three kilometers ($1/2$ to 2 miles) wide, which accommodates one or more utility facilities, such as pipelines. Connects the site (source) to a delivery point.

CPUC: California Public Utilities Commission. The state agency created by a constitutional amendment in 1911 to regulate the rates and services of more than 1,500 privately owned utilities and 20,000 transportation companies operating in California.

Design Life: The expected duration or useful life of the pipeline. Often up to 40 years, but may be the projected life of the supply or market.

Digital Orthophoto: A computer-rectified aerial photograph. A raster image of ground features in their true map position. Unlike a photo enlargement, orthophotos have a consistent scale; as such, they provide accurate distances, angles, positions, and areas. They are often used as an accurate hard-copy backdrop for creating overlays of a wide variety of information, such as property data and land use cover. Digital orthophotos are frequently used to create an accurate GIS base map. *See also GIS.*

DOE: Department of Energy.

DOI: Department of the Interior. The BLM is an agency within the DOI.

DOT: Department of Transportation. The federal agency responsible for regulating the safety of natural gas and liquid pipelines—both inter- and intrastate. The DOT publishes the Pipeline Safety Regulations Manual, Parts 190–199.

EIR: Environmental Impact Report. A detailed statement required by a state utilities' regulatory commission (PUC) describing and analyzing the significant effects of a project, and discussing ways to mitigate or avoid any adverse effects. Also DEIR and FEIR—draft and final environmental impact reports.

EIS: Environmental Impact Statement. A document prepared pursuant to the National Environmental Policy Act (NEPA). The federal agencies' equivalent of the EIR. Also DEIS and FEIS—draft and final environmental impact statements.

Encroachments: Unauthorized land uses or structures occurring within a right-of-way.

Environmental Assessment (EA): Or Proponent's Environmental Assessment (PEA). An environmental document, usually prepared by, or for, the project proponent, and which may form the basis for a negative declaration, or EIS/EIR. The EIS or EIR will be prepared by the lead agency.

EPA: Environmental Protection Agency. A federal agency created to implement policies and actions that aim to protect human health and the environment.

Fault (Geologic): Break in the earth's crust along which parallel slippage of adjacent material has occurred at some point in the past.

GLOSSARY OF TERMS

Feasible: Capable of being accomplished successfully within a reasonable period of time, taking into account economic, environmental, legal, social, and technological factors.

FERC: Federal Energy Regulatory Commission. The FERC has the responsibility of regulating pipelines.

FHWA: Federal Highway Administration.

FTC: Federal Trade Commission. The FTC has the responsibility of regulating pipelines.

GIS: Geographic Information System. Technology that applies spatial reference to derived information by combining database attributes and digital mapping. Computer hardware and software systems display base map backgrounds with digital coordinates. In this mode, a variety of database information is linked together in a common format. Different layers of information are integrated in a spatial frame of reference or geographic locale to assess mutual impact or relationships. Derived data may include assessment of, say, land development versus environmental impacts, ownership records versus corridors and tax authorities, and many other engineering, marketing, environmental, or land use concerns.

GPS: Global Positioning System. A system of 24 satellites provided by the Department of Defense that provides a user with geodetic or geographic coordinates 24 hours a day in any kind of weather, anywhere in the world. Depending on the type (can be as small as handheld), number of receivers used, and mode of operation, accuracy can range from a few millimeters to 100 meters.

Ground Profile: The vertical elevations plotted along the alignment of the pipeline centerline.

Groundwater: Water within the earth that supplies wells and springs.

Groundwater Table: Fluctuating level of groundwater within the earth; high during rainy season, low during dry season.

Habitat: The place or type of site where a plant or animal is found, either naturally or commonly.

Interstate: Existing between, or connecting, two or more states. Pipelines are regulated by the FERC and the FTC.

Intrastate: Existing within the boundaries of one state. Pipelines may be regulated by a state public utilities commission or regulatory body.

Lead Agency: The federal, state, or local agency charged with preparing an EIS or EIR. Often the agency whose jurisdiction is most affected by the pipeline route.

Location Engineer: In this manual, this term is equal to a routing engineer (routing engineer is used here to differentiate it from those engineers who locate existing lines).

Long-Term Impact: The effect or damage caused by the construction of the pipeline that extends after construction. Examples are a permanent access road, or removal of rows of trees that cannot be replanted after the pipe is installed.

Mitigation: Measures instituted to minimize loss of land or land use; measures to reduce environmental impact or hazards to pipeline.

NEPA: National Environmental Policy Act. A federal act, enacted in 1969.

NTSB: National Transportation Safety Board. Federal agency responsible for overseeing safety in the pipeline industry.

Parallel or Common Corridor: A corridor that follows and/or overlaps existing facilities.

PIP: Public Involvement Program. The program developed and implemented by the project proponent in order to provide organized and consistent project information dissemination; also, ensures public participation in the decision-making process.

Public Involvement: A process by which the project proponent consults and obtains comments from interested and affected individuals, organizations, agencies, and governmental entities when planning a new major utility.

QL: Quality level of existing underground utility information that is depicted on the plans or construction document.

Right-of-Way: Often abbreviated as ROW or R/W. A specific strip of land within a route, usually from 5 to 30 meters (15 to 100 feet) wide, within which the user has the authority to construct and maintain utility facilities, such as pipelines, under a written agreement with the property owner.

Rod: An old unit of measurement equal to $16^1/_2$ feet.

Route: A strip of land of varying width, often 300 to 400 meters (1000 feet to $1/_4$ mile) wide, within a corridor, and in which a right-of-way for a utility facility, such as a pipeline, could be located.

Route Criteria: Or Evaluation Criteria. Commonly divided into types of engineering factors that affect design and construction of the pipeline, and existing and future environmental issues—the bases on which the pipeline alignment will selected. Examples include: length of route; number of water crossings, and wildlife habitat.

Route Methodology: Or Evaluation Methodology. Steps used to identify and define evaluation criteria, collection and inventory of available data, review of route data for all alternatives, and comparison of routes with summary.

Routing Engineer: The person responsible for that part of the project that has to do with the selection of the final pipeline alignment. Sometimes this person is known as the location or right-of-way engineer. The routing engineer on smaller projects, or with different companies, could also be the design engineer or project manager.

Scoping: A process used to solicit input from federal, state, and local agencies, as well as interested citizens and organizations, on the issues that they believe should be addressed in the environmental report prepared for the project.

Short-Term Impact: Effects on the environment or land that last for the duration of construction, or for a short time after. Examples are temporary realignment of a ditch, or taking crops out of production for only one growing season.

SLC: State Lands Commission. Often the lead agency for preparation of an EIS or EIR.

Staging Area: The site headquarters where employees start and end each day. These sites also serve as materials and equipment storage areas.

Study Area: The geographical area of a region established for the siting or modification of proposed utility facilities such as pipelines.

Study Corridor: A strip of land that connects the site, or source, to a delivery point, and which will be compared with issues arising in other selected corridors.

SUE: Subsurface Utility Engineering.

USACOE: United States Army Corps of Engineers. May be the lead agency for an EIS if the pipeline crosses its jurisdiction. The USACOE has jurisdiction over all navigable waters within the USA. In actuality, the legal definition of navigable waters has expanded to include marshes, swamps, and diked lands, even if not navigable. Any entity wishing to locate a pipeline, excavate, or discharge dredged material within the Corps' jurisdiction must obtain a 404 permit.

USFS: United States Forest Service. May be the lead agency for an EIS if the pipeline crosses its jurisdiction.

USFWS: United States Fish and Wildlife Service. May be the lead agency for an EIS if the pipeline crosses its jurisdiction.

USGS: United States Geological Survey. An agency of the federal government charged with the preparation of topographical maps for the entire country. The term "USGS Quad" refers to the actual map; for study purposes this is usually the $7\frac{1}{2}'$ quad at a scale of 1:24,000.

Water Conveyance Facilities: Ditches, canals, pipelines, and other such means of moving water for agricultural or industrial use.

Williamson Act Contract: A contract entered into between a landowner and a county, whereby the landowner agrees to maintain a certain parcel in agricultural use for a fixed period of time, and the county agrees to contrive to tax that parcel at agricultural rates, regardless of the parcel's highest and best use. The Williamson Act was intended to curtail urban sprawl into prairie agricultural land that was being driven by counties taxing that agricultural land as residential or commercial land, which may have been the highest and best use, based on adjacent development.

BIBLIOGRAPHY AND REFERENCES

ASCE (1978). Manuals and reports on engineering practice 34, *Definitions of Surveying and Associated Terms*. Joint Committee of American Society of Civil Engineering and American Congress on Surveying and Mapping, New York.

ASCE (1985). Manuals and reports on engineering practice 64. *Engineering Surveying Manual*. American Society of Civil Engineering, New York.

ASCE (1981). Manuals and reports on engineering practice 75. *Right-Of-Way Surveying Guide*. New York.

ASCE (1995). "Advances in underground pipeline engineering." *Second International Conference: Subsurface Utility Engineering: Upgrading the Quality of Utility Information*, June.

ASCE Pipeline Division (1981). "Joint usage of utility and transportation corridors." *Proceedings of Conference*, New York, September 24–25.

ASCE Pipeline Division (1992). *Pressure Pipeline Design for Water and Wastewater*, Section 7—Pipeline Economics. New York.

Anspach, James H. (1995). "Subsurface utility engineering: standards for the depiction of subsurface utility quality levels within geographic information systems." *Proceedings of the Geographic Information Systems for Transportation Symposium (GIS-T)*, AASHTO, April.

Chamblin, Caroline (1994). "Integrating public concerns into technical decisions." *Civil Engineering News*, September, Kennesaw, GA.

Connor, Desmond M. (1988). "Breaking through the "NIMBY" syndrome." *ASCE Civil Engineering J.*, December, 69–71.

Day, Nicholas B. (1983). "Joint use of utility and transportation corridors." *National Development J.*, January/February.

Day, Nicholas B. (1989). "Pipeline safety and right-of-way mapping." *RICS Land & Minerals Surveying*, January, London.

FHWA (1995). *Subsurface Utility Engineering: A Proven Solution*. Film.

National Science Foundation (1993). Advances in site investigation and detection. *Research Needs in Automated Excavation and Material Handling in the Field*. April.

Stein, Debra (1992). *Winning Community Support for Land Use Projects*. ULI—the Urban Land Institute, Washington, DC.

Transportation Research Board, National Research Council (1988). *Pipelines and Public Safety: Damage Prevention, Land Use, and Emergency Preparedness.* Special Report 219, Washington, DC.

US Department of the Interior, Bureau of Land Management (1992). *Draft Environmental Impact Statement on the Proposed Idaho Power Company 500kV Transmission Line, the Southwest Intertie Project.* BLM, June, Burley, ID.

US Department of Transportation, Federal Highway Administration, Office of Engineering (1994). *Subsurface Utility Engineering—SUE Handbook.* October.

US Department of Transportation, Federal Highway Administration, Office of Engineering (1995). *Program* Guide Utility Adjustments and Accommodation on Federal-Aid Highway Project. FHWA-PD-95-029, July.

US Department of Transportation, Research and Special Programs Administration. *Pipeline Safety Regulations.* Revised periodically.

Wildenradt, Wayne C. (1983). "Changing economic factors affect pipeline design variables." *Pipeline & Gas J.,* July.

INDEX

Above-ground pipelines 26
Access roads; defined 85; existing *vs.* constructed 11, 14
Acquisition agents 69
Aerial photography 7, 31, 69
Aesthetic considerations 65
Alignment; defined 5, 85; selection 35, 36
Alternative corridors; defined 31; selection 29–31
Ancillary facilities; site considerations 35, 37
Angle point 85
Assessor's maps 7–8

Bureau of Land Management maps 7

California Environmental Quality Act (CEQA) 7, 61, 85
Capable corridors; defined 29; identification 30–31, 32
Compressor stations *see* Station facilities
Constraint areas; defined 85; mapping 29–30
Construction activities; backfilling 79; cleanup and restoration 79; clearing of work strip 73–74, 76–77; installing pipe 77–78; mobilization 73; spoil disposal 77; strength testing 79; trenching 77
Corridors; analysis of 31, 33, 63–64; defined 5, 86
Crossings 11, 15–18, 20, 22; crossing angle 15

DC power sources 40
Delivery points 10
Department of Transportation (DOT) 86
Design life 86
Digital orthophoto 86
Draft Environmental Impact Statement (DEIS) 61

Easements; pre-existing 8, 67–68; *vs.* full fee 68
Economic considerations; annual operating expenses 82; cost escalation rate 82–83; cost of building 82; economic climate 83; environmental costs 58; system owner's perception 83; value of money 83
Encroachments 86
Entry agreements 67
Environmental Assessment (EA); defined 86; outline for 62–63; requirement for 61
Environmental considerations; cost 58; and entry agreements 67; impact categories 64–65; project team composition 6; resource categories 62; scoping 61; sensitivity analysis 63; water crossings 11, 15–17, 20, 22; wetlands 15
Environmental Impact Report (EIR); defined 86; requirement for 61; reviewing previous 48
Environmental Impact Statement (EIR); defined 86; requirement for 61; reviewing previous 48

Environmental Protection Agency (EPA) 86

Fault (geologic) 86
Federal Energy Regulatory Commission (FERC) 87
Federal Land Policy and Management Act (FLPMA) 63
Federal Trade Commission (FTC) 87
Forests 23
Franchise use 11, 69–70

Geographic information system (GIS); and corridor studies 63; defined 87
Geological hazards 25–26
Global positioning system (GPS) 87
Ground profile 86
Groundwater 86
Groundwater table 86

Habitat; defined 86; environmental impacts 62, 64–65
Horizontal directional drilling (HDD) 16, 20

Incident mitigation 40, 42
Irrigation routes 20–21

Lakes 22–23
Land ownership maps *see* Assessor's maps
Land rights 67–71; assessor's maps 7–8; damage payments 68; easements *vs.* full fee 68; eminent domain 71; entry agreements 67; franchise use 69–70; landowner relations 70–71; Native American reservations 71
Land use compatibility 16, 21–23
Lead agency; defined 86; responsibilities 61
Location engineer *see* Routing engineer
Long-term impact; defined 87–88; examples 64–65

Maintenance; patrolling pipelines 79
Maps; assessor's 7–8; topographic 7, 30–31
Mineral deposits 23

Mitigation; defined 88; post-construction 79
Mobilization 73

National Environmental Policy Act (NEPA) 7, 61, 88
National Park Service (NPS) 61
National Transportation Safety Board (NTSB) 88
NIMBY (not in my back yard) syndrome 48, 52

Opponents, handling of 51–52, 54
Orchards 16, 22

Pipe diameter 10, 81
Pipe material 10
Political considerations 6, 31, 35; equity/empathy/effort 51; public involvement programs 50–55; typical opponents' concerns 51
Power lines 40
Product considerations 9
Project management; organization chart 3; planning criteria 6–7; sample timeline 2; study process 59–60; team members 9; team relationships 4
Public involvement program (PIP) 50–55
Public meetings; follow-up 55; format 53; notification 53–54; schedule and agenda 54–55
Public safety 39–40
Pump stations *see* Station facilities

Regulatory issues 47–50; *see also* Environmental considerations; agencies 49–50; agencies involved 61
Reservoirs 22–23, 24
Residential areas 21–22
Rice fields 41
Right-of-way (ROW); acquisition 67–71; defined 5, 88; width guidelines 13
Roads; access, defined 85; crossings 11, 16, 18; parallel pipeline routing 11, 14
Route; defined 5, 88

Route identification 34; beneficial condition checklist 27; negative condition checklist 27–28; public involvement 31, 35
Route matrix 63
Routing engineer; defined 88; knowledge 6; qualities 6; responsibilities 5, 47

Sabotage 42
Safety; incident mitigation 40, 42; lifeline safety 40; public safety 39–40; sabotage 42; subsurface utility impacts 43–45
Scoping 61, 88–89
Seismic instability 25–26
Sensitivity analysis 63
Short-term impact 89
Sideslope terrain 25
Soil conditions 25
Soil contamination; pre-existing 22
Spoil disposal 77
Staging area 73, 89
State Lands Commission (SLC) 89
Station facilities; and pipe diameter 81; site considerations 35, 37

Strength testing 79
Study area 89
Study corridors 31, 33, 63–64, 89
Subsurface utility engineering 43–45

Terminal points 10
Terrain considerations 23, 25
Trenchless construction 15–16

United States Army Corps of Engineers (USACOE) 89
United States Geological Survey (USGS) 89; maps 7
Utilities, underground; engineering principles 43; information quality 43, 45; shared franchise use 70

Visual impacts 65

Water crossings 11, 15–17, 19, 22
Wetland areas 15
Williamson Act contract 89
Work strip; clearing 73, 76–77; guidelines 13; width 10